DATE DUE

••21••
ASTOUNDING SCIENCE QUIZZES

21 ASTOUNDING SCIENCE QUIZZES

Grace Marmor Spruch
and Larry Spruch

Cartoons by Nurit

BARNES & NOBLE BOOKS
A DIVISION OF HARPER & ROW, PUBLISHERS
New York, Cambridge, Philadelphia, San Francisco,
London, Mexico City, São Paulo, Sydney

Grateful acknowledgment is made for permission to reprint:

Cartoons on pages 42, 45, 50–51, 70–71, 73 from *No Comment* by Nurit Karlin. Copyright © 1978 by Nurit Karlin (New York: Charles Scribner's Sons, 1978). Reproduced with the permission of Charles Scribner's Sons.

Cartoons on pages 33, 64, 149 originally appeared in *The New Yorker.* Copyright © 1974, 1975, 1978, respectively, by The New Yorker Magazine, Inc. Reprinted by permission.

Cartoons on pages 20, 100–101 originally appeared in *Audubon* magazine.

"Do You Have a Fine 14th- or 20th-Century Mind?" originally appeared in *The New York Times,* August 2, 1976. © 1976 by The New York Times Company. Reprinted by permission.

Since this page cannot legibly accommodate all photo credits, they may be found after page 149.

 For information address Harper & Row, Publishers, Inc., 10 East 53rd Street, New York, N.Y. 10022. Published simultaneously in Canada by Fitzhenry & Whiteside Limited, Toronto.

FIRST EDITION

Designer: Robin Malkin

ISBN 0-06-463550-3

82 83 84 85 86 10 9 8 7 6 5 4 3 2 1

To parents who struggled
so that children could get
the education they missed

Contents

Preface

Science is all around us, from space probes through recombinant DNA down to quarks. Yet direct involvement with science by educated nonscientists is not as great as it was, say, in Jefferson's time.

In Jefferson's day, education was for the few. But those few did a lot of puttering with pieces of apparatus and peering through optical tubes. Jefferson, the gentleman farmer, was a gentleman scientist as well.

Today, education is for the many. More people know a little about science. But cost and complexity have taken science out of the hands of the putterers and put it into the laboratories of the professionals. The fields in which amateurs can "do science" are few. In some branches of astronomy, amateurs have traditionally made discoveries; discovering comets seems to be a specialty of airplane pilots and Japanese on the ground. But what amateur can compete with a radiotelescope or an accelerator one mile across? Today, nonscientists read about science. Science has taken on aspects of a spectator sport.

And just when the amateur has been squeezed out, the game is at its most exciting. Newton could seek the form of the planets' orbits about the sun; we can ask how the planets were *formed.* We can ask questions that may be the deepest that can be asked. Newton could ask them too, of course, but for us the answers may be within reach. What is the origin of the universe? What is the origin of life? What is the *nature* of life? And is there life elsewhere? Will the universe collapse down to a point some day? If so, will it start up again? When these are the issues science confronts (and when the government constantly makes decisions of great importance—and expense—which relate to science and technology), it is no wonder that there is current interest in science; the wonder is that there isn't *more* interest.

There has indeed been a recent wave of interest in science, as evidenced by the appearance of popular magazines on science and a host of TV programs, one series boasting a former newsman who lifted anchor to lend his name to the universe and his insights to natural, rather than people-made, catastrophes. The wave of interest hasn't flooded many science classrooms, however; the "in" subject at this moment is business, which usurped pre-eminence from pre-med. Interest in science is apparently satisfied extracurricularly.

This collection of quizzes joins the magazines and TV programs in aiming to bring science to the interested person on an accessible level. However, while magazines and TV present science in the manner of a scientific paper but with the science watered down, the approach here is to make the spectator participate a bit more in the sport. And, since the specialist in one area of science is often the spectator of another, these quizzes are for anyone who wants to check how well he is keeping up with science. Not ordinary classroom quizzes, they make frequent forays into literature, art, and music; in that sense they are interdisciplinary and, we hope, will help narrow the gap between the two cultures.

The quizzes began with our reading an article in *The New York Times* relating how American youth had fared badly on a history test. The *Times* threw down the gauntlet by printing the test. As we tackled the questions, we began to wonder how American youth would fare on a science test. Wondering led to writing. We constructed a science quiz and, although we sent it to the *Times,* we really didn't expect them to print it. We had put a respectable amount of effort into the quiz's preparation, but less than if we had been certain of publication and nowhere near that involved in writing a scientific paper (which is usually read by only a handful of experts); and so, when the *Times* did publish it—on the Op-Ed page—the effort-to-reader ratio was, as mathematicians would say, "vanishingly small." We were reminded of movie mogol Adolph Zukor's remark, on his one-hundredth birthday, that, if he had known he would live so long, he would have taken better care of himself. Unlike Zukor, we were given a second chance. The chance to take better care came from Robert Ubell, then editor of *The Sciences,* the magazine of the New York Academy of Sciences, who liked our quiz and asked us to write one for each issue of his magazine. The quizzes collected here represent a part of the product of a very pleasant association with *The Sciences.* All but the first quiz appeared there originally. The first is essentially the one printed by the *Times,* bearing the title supplied by Howard Goldberg, an editor on the Op-Ed page, an adaptation of a line from *Jacobowsky and the Colonel.* When given the opportunity to put greater effort into the *Times*'s quiz, we ended up making only minor modifications. (There may be a lesson in this about life.) While all the quizzes have been altered only slightly from their original versions, the answers have been expanded considerably. Those in *The Sciences* consisted almost exclusively of a letter indicating the correct choice. For this book, we have added bits of information about correct and incorrect answers alike, for sometimes the incorrect answers are more interesting than the correct ones.

We suggest you complete an entire quiz before looking up answers, because an early answer may give away one to come. When the quizzes were collected, we removed most of the repetitions in subject and alternative choices from one quiz to another. Those that remain are not the result of oversight but can be attributed to that disorder of pedagogues, the chronic belief that, when it comes to learning, a little repetition can be a good thing. Optimism is also endemic in pedagogues. We hope the quizzes will prove a source of both learning and enjoyment.

We were fortunate in being able to obtain cartoons by Nurit, whose way of looking at things we found so nuttily right.

Throughout, we were blessed with good friends and kind

colleagues whom we were able to exploit. The list of those who served nobly and well in debugging a great many quizzes is headed by Professors Lawrence A. Bornstein and Englebert L. Schucking of New York University, and Dr. Richard E. Silverman. Tapped when standard references proved wanting were Doctors Israel Wilenitz and Joan Silverman. Others, drafted for the occasion when a quiz overlapped their areas of expertise, included a Rutgers contingent, Doctors Michael Aissen, Alan D. Bernstein, James R. Freeman, Yuan Li, Don Salisbury, and Vincent Santarelli; a New York University contingent, Paul Gans, Nicholas Geocintov, Michael Goodman, Jerome K. Percus, V. T. Rajan, Robert W. Richardson, and Edward J. Robinson; and a group with affiliations ranging from California to Long Island, Canada to Louisiana: Evelyn Berezin, Richard Stephen Berry, Sidney Borowitz, Jeffrey Goldstone, Madeleine and Morton Hamermesh, Joseph B. Keller, Derek A. L. Paul, A. R. P. Rau, the late Sol Rubinow, Malvin A. Ruderman, William C. Saslaw, and Robin Shakeshaft.

If we are informed of any bugs that still remain, we will "pass the bug" to the above.

Special thanks go to Jeanne Flagg, this book's editor, for constructive editing in general and masterly ambiguity-detecting in particular.

While most of the quizzes were written while we were at Rutgers, Newark, and New York University, we wish also to acknowledge, as scenes of inspiration, the Aspen Center for Physics and Harvard University, and for stimulating atmosphere during final manuscript adjustments, the Institute for Advanced Study and Princeton University.

••21••
ASTOUNDING SCIENCE QUIZZES

1

Do You Have a Fine 14th- or 20th-Century Mind?

1. The ratio of the kilometer to the mile is roughly a) 1 to 10 b) 5 to 8 c) 8 to 5 d) 2 to 1.
2. Water freezes at a) zero degrees Fahrenheit b) 32 degrees Fahrenheit c) 100 degrees Celsius d) absolute zero.
3. A lunar eclipse can occur only when a) the Earth is between the sun and the moon b) the moon is between the Earth and the sun c) the sun is between the moon and the Earth d) there is a new moon.
4. The conservation-of-energy principle refers to the fact that a) it is essential not to waste natural gas and oil, for these are limited in supply b) solar heating makes use of the sun's energy, which would otherwise be wasted c) energy can be neither created nor destroyed d) nuclear power plants recycle spent fuel.
5. The splitting of an atomic nucleus into two large fragments and several smaller ones is known as a) fusion b) alpha decay c) fission d) thermonuclear energy.
6. Atoms are believed to be composed of a) protons, neutrons, and electrons b) protons and electrons c) positrons, neutrinos, and electrons d) protons and antiprotons.
7. The period of revolution of the moon about the Earth is approximately a) one year b) one month c) one day d) one hour.
8. Identify the nonastronomical objects: a) white giants

and green dwarfs b) white dwarfs and black holes c) quasars and supernovae d) neutron stars and galaxies.

9. An outstanding Soviet dissident who is a physicist is a) Rostropovitch b) Sakharov c) Mendeleev d) Baryshnikov.
10. The Pythagorean Theorem states that a) in any triangle the square of the longest side equals the sum of the squares of the other sides b) in any triangle the square of the longest side equals the square of the sum of the other sides c) in a right triangle the square of the hypotenuse equals the sum of the squares of the other sides d) in an isosceles triangle the third side equals the sum of the two equal sides.
11. One type of radioactivity involves a) gamma rays b) sunspots c) pulsars d) magnetic fields.
12. Thirteen billion years corresponds most closely to the presumed a) age of the universe b) age of the Earth c) time since the dinosaurs were on Earth d) time man has been on Earth.
13. Helium was first discovered a) in mines b) in the depths of the ocean c) on the moon d) on the sun.

14. A planet that is never visible to the naked eye is a) Mercury b) Venus c) Mars d) Neptune.

15. The chain reaction that forms the basis of the atomic bomb was first achieved by a group directed by a) Albert Einstein b) Niels Bohr c) Edward Teller d) Enrico Fermi.

16. The gravitational force between two spherical objects is known to be inversely proportional to the square of the distance between their centers. If that distance is made three times larger, the gravitational force will be a) ⅓ as large b) ⅑ as large c) half as large d) three times larger.

17. Who did not make fundamental contributions to the science of electricity? a) Charles Coulomb b) Michael Faraday c) Benjamin Franklin d) Isaac Newton.

18. The Big Bang is related to a) the hydrogen bomb b) the maximum noise level in an amplifier c) a theory of the origin of the universe d) supersonic aircraft.

19. Nuclear physics does not deal with a) alpha particles b) beta rays c) deuterons d) deoxyribonucleic acid.

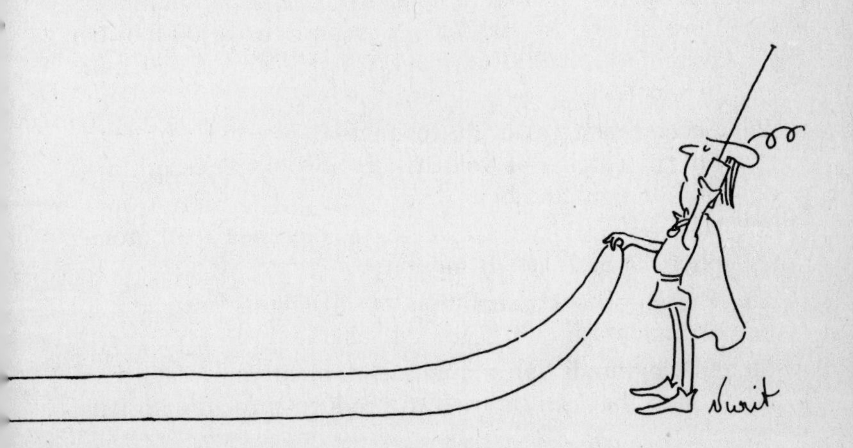

20. Identify the incorrect statement. Transmutation of the elements a) was a goal of the alchemists b) occurs in ordinary chemical reactions c) was first achieved by Ernest Rutherford d) occurs in some nuclear reactions.
21. Radiocarbon dating is a technique by which a) persons who might get along well together are identified by computer b) the fading of carbon copies is used to tell the age of documents c) the age of art objects is measured d) the length of time a patient has had cancer is determined.
22. A laser is not a) a source of light that can be focused to a tiny area b) a device conceived by Jules Verne for propelling a man to the moon c) employed in some delicate eye operations d) a device that was used to measure the distance to the moon.
23. Light a) can travel in a vacuum b) can travel at infinite speed c) always travels in perfectly straight lines d) cannot travel through solid objects.
24. A rocket moves because a) its shape permits air to support it b) it has exceptionally powerful propellers c) it weighs less than the air it displaces d) there is a reaction to the gases it exhausts.
25. The speed of sound in air under standard conditions is most nearly a) 10 feet per second b) 1,000 feet per second c) 10,000 feet per second d) 186,000 miles per second.
26. Acceleration a) is the change in velocity b) is the rate of change of velocity c) always increases d) is the force on an object.
27. Newton's three laws relate to a) electricity b) atomic physics c) heat d) motion.
28. There is no conservation of a) angular momentum b) momentum c) force d) charge.
29. A hologram is a) a rapid means of communication b) a slide that can be used to produce three-dimensional

images c) an atom smasher d) a future mode of transportation.

30. The "Red Planet" is a) Saturn b) Venus c) Sputnik d) Mars.

31. The ancient Greek scientist one associates with an atomic theory is a) Archimedes b) Pythagoras c) Eureka d) Democritus.

32. A half-life is a) a molecule that cannot be classed as definitely organic or definitely inorganic b) half the average life expectancy of a group of people c) the time for half a given amount of radioactive material to decay d) the radiation dose that will be lethal to half the subjects in an experiment.

33. Give the proper order of the names Archimedes, Copernicus, Einstein and Galileo so that they correspond to the order of these statements:
—Was the first to view the moons of Jupiter through a telescope.
—Showed the equivalence of mass and energy.
—Stated that a floating body displaces a volume of water the weight of which equals the weight of the body.
—Stated that the sun, rather than the Earth, is at the center of the solar system.
a) Archimedes, Einstein, Galileo, Copernicus
b) Copernicus, Einstein, Archimedes, Galileo
c) Copernicus, Archimedes, Galileo, Einstein
d) Galileo, Einstein, Archimedes, Copernicus.

34. A topic not likely to arise in SALT talks is a) NaCl b) ICBM c) MIRV d) U-235.

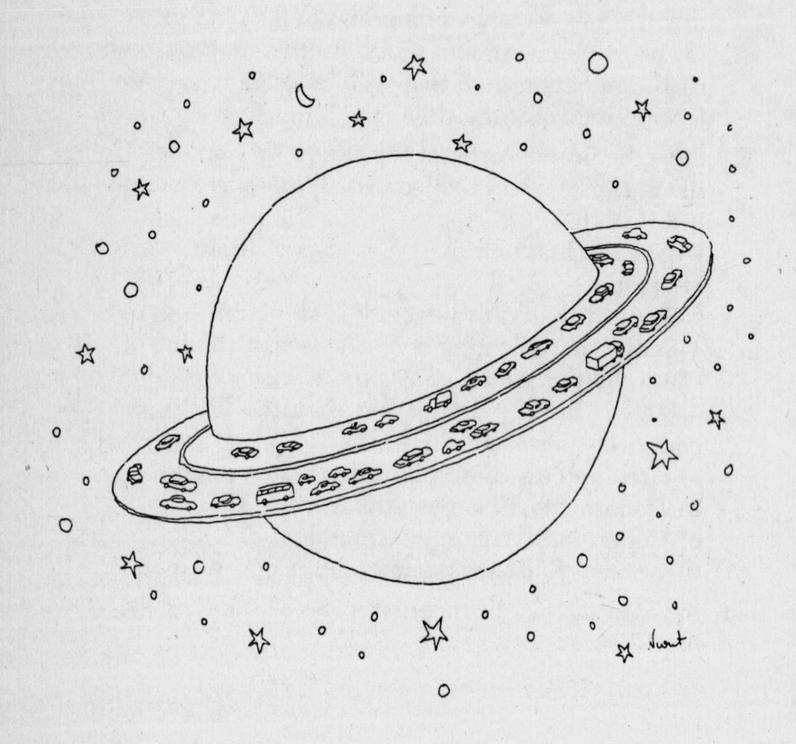

•• 2 ••

Far Out

1. In the mid 70s Viking expedition to Mars—roughly 200 million miles away at the time—the flight was controlled from Earth only until the last half hour, when the lander descended from an orbit around Mars to the designated site on the surface. Why? **a)** The landing was on the side of Mars away from Earth and therefore could not be controlled from Earth. **b)** The lander was passing through Mars's atmosphere, which absorbs radio signals. **c)** The transit time of signals from Viking to Earth and back was greater than the landing time. **d)** The controllers at the Earth space center were exhausted by then.
2. The energy we receive from the sun is produced primarily through the process of **a)** fission **b)** fusion **c)** gravitational collapse **d)** oxidation.
3. A quasar is *not* **a)** an astronomical entity discovered within the last 25 years **b)** a quasi-stellar object **c)** the name of a TV set **d)** a U.S. Air Force satellite.
4. An astronomical object found in the Bayeux tapestry (depicting the Norman conquest of England) and also associated with Mark Twain is **a)** Halley's comet **b)** the Andromeda galaxy **c)** the Crab nebula **d)** the North Star.
5. At sunset, the sun appears red. The reason is: **a)** "red hot" is cooler than "white hot" and, at the end of the day, the sun's temperature goes down **b)** a psychological

effect—the sun is low on the horizon and seems to be near buildings or fields, the colors of which make the sun appear red **c)** at sunset, the sun's light has more atmosphere to traverse, atmosphere which scatters blue light more than red, leaving the red to give the sun its appearance **d)** there is less light at sunset and, with less light, the sun appears darker.

6. The number of mythological figures who have *both* planets and chemical elements named *directly* after them is **a)** 2 **b)** 3 **c)** 4 **d)** 10.
7. Which is *not* associated with astronomy or space? **a)** UHURU **b)** Cygnus X-I **c)** ETAOINSHRDLU **d)** Hercules X-1.
8. Black holes **a)** have definitely been observed **b)** can never be detected **c)** are bodies from which no light can escape **d)** are gaps in the solar spectrum.
9. The substance that is most abundant in the universe is **a)** hydrogen **b)** oxygen **c)** nitrogen **d)** alcohol.
10. Einstein's *special* theory of relativity, containing his famous formula for the equivalence of mass and energy, $E = mc^2$, has been confirmed in innumerable experiments. One of the few experiments that tested his *general* theory of relativity involved measuring **a)** the bending of a distant star's light rays that grazed the sun **b)** the decrease in the radii of the orbits of the planets with time **c)** the decrease in intensity of the sun's light with time **d)** the change in the period of revolution of certain comets.
11. The temperature at the *surface* of the sun is approximately **a)** 500°K **b)** 5000°K **c)** 500,000°K **d)** 5,000,000°K.
12. The temperature at the *center* of the sun is believed to be about **a)** 10 thousand °K **b)** 10 million °K **c)** 10 billion °K **d)** 10 trillion °K.
13. The longest period of revolution about the sun belongs

to the planet a) Mercury b) Jupiter c) Uranus d) Pluto.

14. In which constellation is there a "star" that appears double to someone with very good eyesight? (In fact, centuries ago, the ability to see this "star" as two was used as a test of visual acuity.) a) the Big Dipper b) the Little Dipper c) Orion d) the Southern Cross.

15. The aurora borealis a) is light reflected from the polar ice cap b) was discovered by Professor Arturo Borealis c) consists of a shower of meteors d) originates in interaction of the atmosphere with incoming charged particles.

16. The universe is believed to be expanding according to Hubble's law, named for Edwin Hubble, the American astronomer who proposed it. The law, valid for almost any observer, including one on Earth, states that $v = Hd$, where v is the speed with which a galaxy recedes from us, d is the distance from us, and H is the Hubble constant, 1/(10 billion years). One interpretation of the law is, *very* roughly, that all the galaxies were created at a single point with a range of speeds; a galaxy with a speed v has been traveling since the creation of the universe T years ago and has gone a distance $d = vT$ from us. The value of T, the age of the universe, is therefore of the order of a) 100 million years b) 1 billion years c) 10 billion years d) 100 billion years.

17. Radio astronomy involves a) using radio waves to direct space vehicles to the planets b) studying radio waves from astronomical objects c) studying the spectra of radioactive materials on the planets d) bouncing radio waves off distant stars.

18. The apparent shift in position of two stars with the movement of the Earth in its orbit (akin to the shift of the needle on a car's speedometer when the passenger viewing it leans from his seat toward the steering wheel) is

known as **a)** parallax **b)** parsecs **c)** perturbation **d)** parturition.

19. Which planet is associated with each of these symbols?

♆ ♂ ♀ ⊕

20. The heavenly bodies have inspired not only songs and poems but candy bars as well. Name two.

••3••

Made in U.S.A.

1. The U.S. is one of the last holdouts against the metric system. A standard set of units for length, mass, and time in the metric system is a) foot, pound, second b) meter, kilogram, hour c) centipede, gram, second d) meter, kilogram, second.
2. The pound and a metric unit have the following relation: a) 2.2 kilograms weigh one pound b) 1 kilogram weighs 2.2 pounds c) 454 kilograms weigh 1 pound d) 20 shillings weigh 1 pound.
3. Which invention is *not* an American invention? a) radar b) the transistor c) the cotton gin d) the zipper.
4. Which *is* American in origin? a) holography b) photography c) the jet engine d) the phonograph.
5. The inventor of the lightning rod was a) Thomas A. Edison b) Charles P. Steinmetz c) Charles C. Lightning d) Benjamin Franklin.
6. Which common American "ailment" is not known in Europe? a) German measles b) poison ivy c) whooping cough d) migraine headache.
7. A person sometimes credited with the discovery of the Gulf Stream is a) William Beebe b) Admiral Byrd c) Benjamin Franklin d) Jacques Cousteau.
8. Michael Faraday, in England, developed the concept of electromagnetic induction, which plays a fundamental role in electromagnetic theory. The same concept was arrived at independently by the American, Joseph

Henry, who helped organize the National Academy of Sciences. Electromagnetic induction is the fundamental basis of operation of **a)** motors and generators **b)** masers and lasers **c)** transistors **d)** light bulbs.

9. Which is *not* the name of a U.S. national laboratory? **a)** Brookhaven **b)** Oak Ridge **c)** Argonne **d)** Marne.

10. The glass harmonica produces tones in the same way your wet finger does when rubbing the rim of a partly filled glass of water. A type of glass harmonica was invented by **a)** Charles Ives **b)** Stephen Foster **c)** Bora Minnevitch **d)** Benjamin Franklin.

11. The American physicist Henry A. Rowland developed a diffraction grating, which is basically a series of slits used to measure the wavelength of light. One version of the grating consists of lines accurately ruled on a transparent material—a glass ruler, in essence. For a grating to measure the wavelength of light, the distance separating the lines should be comparable to the wavelength. If you are told that the wavelength of yellow light is approximately 5000 Angstrom units and that one Angstrom unit is 10^{-8} cm (one one-hundred-millionth of a centimeter), roughly how many lines per centimeter should be ruled to measure the wavelength of yellow light? **a)** 100 **b)** 10,000 **c)** 1 million **d)** 1 billion.

Elementary Questions

12. Which American has an element named after him? **a)** Ernest O. Lawrence **b)** J. Robert Oppenheimer **c)** Edward Teller **d)** Benjamin Franklin.

13. Which municipality has an element named after it? **a)** Chicago **b)** Reno **c)** Berkeley **d)** Miami.

14. Which state has an element named after it? **a)** Illinois **b)** California **c)** Texas **d)** New York.

15. The element the entire country can claim is **a)** americium **b)** usium **c)** unclesamium **d)** columbium.

16. In one of the great experiments in physics, Robert A. Millikan measured **a)** the mass of the hydrogen atom **b)** the mass of the electron **c)** the ratio of the charge of the electron to its mass **d)** the charge of the electron.

17. Another great experiment was the Michelson-Morley experiment. (A.A. Michelson became the first American scientist to win a Nobel prize.) The experiment **a)** confirmed the existence of photons **b)** established that $E=mc^2$ **c)** showed that the speed of light is constant, independent of the motion of the observer **d)** showed that the speed of light varies with the speed of the person observing it.

Prize Questions

18. Which of the following is a Nobel laureate? **a)** J. Robert Oppenheimer **b)** Edward Teller **c)** James D. Watson **d)** Jonas Salk.

19. Which of these American Nobel laureates in physics was born in the United States? **a)** A. A. Michelson **b)** I. I. Rabi **c)** M. Gell-Mann **d)** T. D. Lee.

20. Which American won Nobel prizes in two different fields? **a)** Linus Pauling **b)** Harold Urey **c)** Glenn Seaborg **d)** Theodore Roosevelt.

21. Which American won two Nobel prizes in physics? **a)** John Bardeen **b)** William B. Shockley **c)** Julian Schwinger **d)** C. N. Yang.

22. Lest you think Nobel prizes come cheaper by the dozen, let us go outside the United States to ask who is the only other person to win two Nobel prizes in science? **a)** W. C. Roentgen **b)** Ernest Rutherford **c)** Marie Curie **d)** Ivan Pavlov.

23. A wire is attached to a battery to form an electrical circuit. In the absence of exact knowledge, two equivalent descriptions of the current in the wire are: positive carriers of electricity travel from the positive plate of the battery to the negative; negative carriers travel from the negative plate of the battery to the positive. Our old friend Benjamin Franklin effectively chose the first in the days before the nature of the carriers was known, and his description persists to this day. Carriers can be either positive or negative, depending upon the situation. In a metal, the carriers are **a)** ions **b)** nuclei **c)** electrons **d)** protons.

24. The signer of the letter to President Franklin D. Roosevelt which led him to set up the atomic bomb project was **a)** J. Robert Oppenheimer **b)** General Leslie R. Groves **c)** Winston Churchill **d)** Albert Einstein.

25. Which famous trial was intimately connected with the theory of evolution? **a)** Scottsboro **b)** Scopes **c)** Snopes **d)** Jukes.

26. Give the field of scientific knowledge associated with the following presidents: Thomas Jefferson, Ulysses S. Grant, Herbert Hoover, Richard M. Nixon.

••4••
Misfits

The objective of this quiz is to determine the odd entry. When four items are compared, it is normally possible to find some similarities—and some differences. In each of the following sets of four answers to the questions posed, only three items belong together.

1. Pick the misfit in time. Who was *not* a Renaissance man? a) Brahe b) Bruno c) Galen d) Galileo.
2. Pick the misfit in place. Who was *not* born in the U.S.? a) Josiah Willard Gibbs b) Joseph Henry c) Count Rumford d) Joseph John Thomson.
3. Pick the odd country. a) the one where the steam engine was invented b) the one where the maser was invented c) the one where the first successful flight in a motor-powered airplane took place d) the one where the cyclotron was invented.
4. Which of the following personages, eminent in one field, did *not*, in addition, have a strong interest in mathematics? a) Eamon de Valera b) Ignace Jan Paderewski c) Ulysses S. Grant d) Lewis Carroll.
5. Which principle would *not* normally be found in an advanced physics text? a) the correspondence principle b) the exclusion principle c) the principle of marginal utility d) the uncertainty principle.

6. Who would have had a hard time getting into the electricians' union (theoretically speaking)? **a)** Andre Marie Ampère **b)** Charles Augustin de Coulomb **c)** Robert Hooke **d)** Alessandro Volta.
7. Which of the following courses would *not* be likely to be offered by a physics or chemistry department? **a)** spectroscopy **b)** holography **c)** crystallography **d)** proctology.
8. Which of the following is *not* a unit of length? **a)** a mil **b)** a light-year **c)** a parsec **d)** a radian.
9. Which unit would *not* be employed in a measurement of area? **a)** a barn **b)** a silo **c)** a hectare **d)** an acre.
10. Which of the following men would have been most likely to have failed an examination in mathematics? **a)** Aeschylus **b)** Archimedes **c)** Euclid **d)** Pythagoras.
11. Which is *not* a designation of a class of numbers? **a)** rational **b)** perfect **c)** prime **d)** choice.
12. Which of the following is *not* a mathematical entity? **a)** a matrix **b)** a vector **c)** a lattice **d)** a specter.
13. Which of the following words is *not* a technical term pertaining to time? **a)** sidereal **b)** solar **c)** synodic **d)** surreal.
14. One of the classic tests of Einstein's general theory of relativity involves measurements of the "aberrant" motion of the planet Mercury. A word that would *not* arise in a discussion of such a test is **a)** perihelion **b)** peristylium **c)** perturbation **d)** precession.

15. Which name is *not* associated with the space program? a) Explorer b) Viking c) Adventurer d) Pioneer.

16. Who among the following was *not* a terrestrial explorer? a) James Cook b) Richard E. Byrd c) Antoine Capon d) Giovanni Caboto.

17. Which is *not* a theory once held and later supplanted? a) the phlogiston theory b) the quantum theory c) the geocentric theory d) the ether theory.

18. Which of the following persons did *not* have his name given to a constant of nature? a) Avogadro b) Boltzmann c) Loschmidt d) Wolfschmidt.

19. Which would *not* be encountered by a nuclear physicist measuring the radiation emitted by radioactive materials? a) alpha particles b) beta particles c) gamma rays d) epsilon rays.

20. Who would have been an unlikely candidate for election to a national physics society? a) Arnold Toynbee b) Ernest Rutherford c) Wolfgang Pauli d) Niels Bohr.

21. Who cannot be credited with having developed a vaccine against a disease? a) Louis Pasteur b) Linus Pauling c) Albert Sabin d) Jonas Salk.

22. Which number is a radical among squares? a) 9^2 b) $191^{1/2}$ c) 225 d) 1024.

23. A word that would *not* arise in an engineering text in a chapter devoted to forces is a) compulsive b) harmonic c) impulsive d) constant.

24. Who could *not* be labeled an inventor? **a)** Fulton **b)** de Kalb **c)** Whitney **d)** Wright.

25. Which of the following has *no* astronomical association? **a)** Perseus **b)** Perstirpes **c)** Pisces **d)** Pleiades.

26. Which word is *not* used in computer terminology? **a)** binary **b)** tertiary **c)** hardware **d)** software.

5

Designer Genes

1. Companies such as Biogen and Genentech engage in what is called **a)** eugenics **b)** biodegrading **c)** genetic engineering **d)** biotechnics.
2. Scandinavians have light skin. One possible explanation is: **a)** The pigment in their skin has been bleached by extensive exposure on nude beaches **b)** It is purely accidental, there is no possible explanation **c)** Lack of pigment is advantageous in regions where there is little sunlight **d)** Light skin is associated with a high intake of milk, which contains oxalic acid, a bleaching agent.
3. The number of cells in the average human adult is about **a)** 50 thousand **b)** 50 million **c)** 50 billion **d)** 50 trillion.
4. Each species has a characteristic number of chromosomes. Human cells contain **a)** 26 pairs **b)** 23 pairs **c)** 46 pairs **d)** about 10 thousand.
5. Cell division is known as **a)** lordosis **b)** mitosis **c)** thrombosis **d)** halitosis.

6. There have been claims that a certain combination of chromosomes could be associated with aggression in male human beings. That combination is **a)** XX **b)** YY **c)** XXY **d)** XYY.
7. If a person has one gene for blue eyes and one for brown, then, according to simple Mendelian laws, that individual will have **a)** one blue eye and one brown eye **b)** brown eyes because the gene for brown is dominant over that for blue **c)** half a chance of having blue eyes and half a chance of having brown **d)** hazel eyes because hazel is an average of blue and brown.
8. A lethal mutation is one that **a)** causes the death of the species **b)** produces a trait that is lethal only to certain members of the species **c)** produces a gene that destroys other genes **d)** brings about death of the individual before the gene producing it can be transmitted to another generation.
9. Both parents have one gene for brown eyes and one for blue. According to simple Mendelian laws, what is the probability that a child will be blue-eyed? **a)** 1 **b)** ½ **c)** ¼ **d)** 0.
10. The name of the charlatan geneticist who thrived during the Stalin era and had a devastating effect on Soviet genetics for years afterward is **a)** Potemkin **b)** Rasputin **c)** Lysenko **d)** Chichikov.
11. If one parent has one gene for brown eyes and one for blue, and the other parent has two genes for blue eyes, what is the probability, according to simple Mendelian laws, that a child will have brown eyes? **a)** 1 **b)** ½ **c)** ¼ **d)** 0.
12. Which statement about sickle-cell anemia is incorrect? **a)** People with two genes for sickle-cell anemia usually die young **b)** People with one gene for sickle-cell anemia are more resistant to malaria than others **c)** Sickle-cell anemia derives its name from the shape of the affected red blood cells **d)** Sickle-cell anemia derives

its name from the emblem of the country of the Soviet scientist who discovered the disease.

13. Winners of the Nobel prize in medicine or physiology for work in genetics include all but **a)** Norman Borlaug **b)** Jacques Monod **c)** Salvator Luria **d)** Maurice Wilkins.

14. If a characteristic that has two versions is determined by only one gene, and if 80 percent of the genes in the gene pool are for the dominant version, what is the probability that an individual will show the recessive version? **a)** 1/25 **b)** 1/5 **c)** 1/4 **d)** 1/2.

15. If an individual has received one gene for blue eyes from each parent, as far as eye color is concerned that individual is **a)** homogamous **b)** homozygous **c)** homogeneous **d)** homeopathic.

16. A somatic mutation is a change that **a)** occurs in an ordinary cell and is not transmitted **b)** occurs in a sex cell and is not transmitted **c)** occurs in *some* cells and not in others **d)** affects the cells associated with sleep.

17. Prior to the 1940s, the only one of the following that could have been responsible for mutations to any extent is **a)** cod liver oil **b)** sunlight **c)** milk **d)** spinach.

18. Down's syndrome **a)** is a chromosomal aberration; the afflicted person has an extra chromosome **b)** is of unknown origin **c)** is a chromosomal aberration; the afflicted person lacks one chromosome **d)** was first reported in Mongolia; hence, it originally was called mongolism.

19. The bases in the DNA molecule do not include **a)** thymine **b)** guanine **c)** cytosine **d)** anodyne.

20. The number of different genes in a human being is estimated to be roughly **a)** ten **b)** ten thousand **c)** ten million **d)** ten billion.

21. Social Darwinism **a)** is a phrase once used to describe Darwin's behavior at social gatherings **b)** constitutes

Darwin's explanation of social behavior and class structure in England, based on principles in the theory of evolution **c)** held that the socially elect classes, in terms of wealth and power, had achieved their position as a result of biological superiority in the struggle for existence **d)** is the theory that, contrary to basic Darwinism, emphasizes cooperation rather than competition.

••6••

Principia Newtonia

There once was a scientist named Newton
Whose equations were not highfalutin.
They governed—though terse—
The entire universe.
And further, they did no pollutin'.

1. The newton is a) a unit of force b) a British coin c) a British car d) a variety of fig.
2. Newton's name was also given to a) a unit of heat b) a type of telescope c) a unit of electricity d) an elementary particle.
3. In Newton's law of universal gravitation, G a) represents the acceleration due to gravity at the surface of the Earth, 9.8 meters/sec^2 b) represents Newton's homage, in the form of a constant, to the great mathematician Gauss c) is the universal gravitational constant d) depends upon whether one is dealing with the Earth, the moon, Mars, etc.
4. If a cannonball and a pebble are dropped from a tower, a) their accelerations will not be about the same b) their final velocities will not be about the same c) the time required for the two to fall will not be about the same d) the pull of gravity on each will not be about the same.

5. An ice skater is slowly turning on one spot, arms extended. If she suddenly pulls her arms in to her sides, a) her speed of rotation will decrease b) her speed of rotation will increase c) her speed of rotation will remain the same d) her motion will stop.
6. Newton was an early scientist in government. He held the position of a) Master of the Mint b) Chancellor of Higher Education c) Science Advisor to the Queen d) Head of the Royal Science Foundation.
7. A problem which Newton solved that is not so well known as some others is a) the meander of a river b) the chemical composition of water c) the physics of musical instruments d) the origin of tides.
8. Conservation laws pertain to quantities that are conserved, or retained, in an event such as a collision, for example. A conservation law that is inherent in Newton's laws of motion is conservation of a) angular velocity b) momentum c) charge d) moment of inertia.
9. The philosophy that gets a direct boost from Newtonian physics is a) idealism b) existentialism c) determinism d) hedonism.
10. Another problem Newton solved was that of the rainbow. He explained it as arising from a) diffraction by the nitrogen molecules which make up about 80% of the atmosphere b) refraction by droplets of moisture in the air c) reflection from the clouds d) a covenant between gravity and light.
11. Throughout his life, Newton devoted a great deal of time to activities other than physics. One was a) playing the violin b) sailing c) womanizing d) explaining biblical chronology.
12. Still another activity Newton engaged in was a) alchemy b) haruspicy c) painting d) writing poetry.
13. Newton's laws enable us to calculate, from first principles, a) the speed of light b) the size of the hydrogen

atom **c)** the force between two charges **d)** the velocity required by a rocket to escape from the Earth.

14. It has been claimed that our system of government was directly influenced by Newtonian physics in that **a)** the first law of motion, the law of inertia, has been likened to strong executive leadership overcoming the inertia of the public **b)** in analogy with the second law of motion, force = mass × acceleration, the democratic process is such that ideas accelerate through the masses and produce a resultant force **c)** the third law, to every action there is an equal and opposite reaction, has been likened to the interplay of the self-interests of different groups **d)** just as each body in the solar system is in equilibrium under opposing forces, the government will remain in equilibrium under the opposing forces of its three branches, the executive, the legislative, and the judicial.

15. Newton (who did not hold with the wave theory of light) discovered that white light is composed of different colors. Which of the following has the longest wavelength? **a)** red **b)** yellow **c)** green **d)** blue.

16. Newton's laws enable one to calculate the mass of any heavenly body that has a satellite. Then, since density = mass/volume, the density can be calculated if one knows the volume from astronomical observations. The mean density of the sun is **a)** much less than that of Earth **b)** much greater than that of Earth **c)** the same as that of Earth **d)** not known.

17. The density of the moon is **a)** much less than that of Earth **b)** much greater than that of Earth **c)** close to but less than that of Earth **d)** close to but less than that of green cheese.

18. Most physicists claim that Newton was the greatest physicist who ever lived. He was both a great experimentalist and a great theorist. Regarding Newton and mathematics, **a)** he flunked math as a schoolboy **b)** many mathematicians claim him as one of the greatest mathematicians who ever lived **c)** he invented partial differential equations **d)** it was a lifelong problem with him and he had to get help in calculus from Leibniz.

••7••

Letters from Greece

If you have never mastered the Greek alphabet, here's an opportunity. Each of the entries below refers to a Greek letter. Where a description might apply to several letters, a unique Greek letter can be obtained by using the additional information provided. In some cases, the relation between entry and letter may be somewhat distant.

1. It makes up the first syllable of the photosensitive pigment also called visual purple, and of a shrub with leathery evergreen leaves. (In a matrix, it sounds like what is not a column.)
2. The chemical symbols for tantalum and uranium spell this letter. (It is also the first syllable of the Latin word for bull.)
3. It means simple *e* to a Greek. (It also sounds somewhat like another Greek letter.)
4. It is a transcendental number. It is also the ratio of the circumference of a circle to its diameter. (Made with apples, it is a particularly American form of dessert.)
5. In a common labeling of axes in three dimensions, it corresponds to *y*. (It sounds like the first part of the word meaning the forms prescribed to be observed in social life, and the first two letters make the suffix meaning *small one.*)
6. It is a light particle emitted by radioactive materials. It is also a symbol for the bulk modulus, and is often used to represent v/c, the ratio of the speed of a particle to the speed of light.

7. It is the nucleus of the helium atom. (It may suggest the name of the sacred river in Coleridge's *Kubla Khan.* It is also the first part of a word for a type of soup.)
8. It is the first syllable of an important word in physics and biology meaning kernel. (It is also the Russian and Yiddish for *Well?*)
9. It is a mathematical function named in honor of Riemann. (The English call the symbol for the capital—if thought of as an English letter—*zed.*)
10. The letter might be suggested by the first syllable of the word for the study of mental processes. (It appears in the middle of the fifth Greek letter.)
11. The symbol for wavelength. The first syllable of this letter is a small domestic animal, and sometimes a term of endearment. The last syllable is the Russian word for *yes.*

12. It is a common symbol for the azimuthal angle that helps position a three-dimensional vector. (It also sounds like the second word in the expression used by Jack's giant.)

13. The symbol represents a singular function, introduced in physics by P. A. M. Dirac. (A shoulder muscle that raises the arm is shaped like the capital letter, and is named for it.)

14. It is, literally, small *o.* (The last syllable starts with the first syllable of a precursor of Homo sapiens.)

15. It is a mathematical function related to the factorial. (A slight mispronunciation gives the name of the physicist author of popular books. It also sounds like the back part of a popular game.)

16. When negatively charged, it is an elementary particle predicted in the early 1960s by Murray Gell-Mann and Yuval Ne'eman. (The last two syllables of this letter, used as a prefix, denote one million. It might also suggest one of Alcott's *Little Women,* had she been Irish.)

17. The capital letter represents the sum in mathematics. (It is the first letter of the honor society that encompasses all the sciences.)

18. Used as a prefix, the symbol denotes one one-millionth. (It is the sound made by a cat.)

19. In some regional accent, the letter might sound like an expression meaning *one should* in the first person singular. If a letter is prefixed, it sounds like a Japanese car. (The letter forms an acronym for Cambridge University's Institute of Theoretical Astronomy.)

20. It might suggest to some the place where one sees plays. (When preceded by a hypo, it becomes somewhat close to imaginary.)

21. It is the first syllable of the word for a fusion and exchange of segments of chromatids during meiosis. (It also forms the beginning of the name of the Windy City.)

22. It contains four successive letters that spell an English word for a tall cylinder used for storing silage. (It might be employed by a child commanding a dog named Silon to stand on its hind legs.)

23. In a common labeling of axes in three dimensions, it is the analog of x. The capital letter consists of three horizontal lines. (It is the first syllable of a percussion instrument consisting of graduated bars.)

24. It is the last letter of a most prestigious undergraduate honor society. (It also sounds like the middle of a word for having "lost one's head.")

8

Two Cultures

1. In recent scientific developments, a word with a literary origin has assumed enormous significance. The literary origin is James Joyce's *Finnegan's Wake.* The word is **a)** snark **b)** sark **c)** narc **d)** quark.
2. The word in the previous question is used in **a)** the theory of elementary particles **b)** the theory of the propagation of electromagnetic waves **c)** the study of the impact of heroin on the central nervous system **d)** the theory of the propagation of gravitational waves.
3. Johann Wolfgang von Goethe was not only a distinguished man of letters; he was interested in a number of other fields, government and science among them. His interest in science was sufficiently great to induce him to formulate a theory of **a)** the origin of the universe **b)** parapsychology **c)** light and color **d)** *son et lumière.*
4. A painter who was a pioneer in the study of anatomy, dissecting cadavers and making sketches of muscles, was

a) Amerighi da Caravaggio b) Rembrandt van Rijn c) Andreas Vesalius d) Leonardo da Vinci.

5. The 1925 Scopes trial dealt with the issue of whether Darwin's theory of evolution could be taught in the public schools of the state of Tennessee. A play about that trial had the title a) *Inherit the Wind* b) *The Wind in the Willows* c) *Trial by Jury* d) *Twelve Angry Men*.

6. A Russian literary work in which one of the main characters is engaged in scientific farming is a) Dostoyevsky's *Notes from the Underground* b) Tolstoy's *Anna Karenina* c) Lysenko's *The Tomato Eaters* d) Gorky's *The Lower Depths*.

7. A piano piece that might come to mind at the sight of a full moon was written by a) Bach b) Beethoven c) Brahms d) Bizet.

8. A second piano piece that might come to mind at the time of a full moon was written by a) Ernest Chausson b) Claude Debussy c) Gabriel Fauré d) Francis Poulenc.

9. Benjamin Franklin is not usually characterized as a literary figure, but his autobiography, a classic, certainly qualifies him as one. His scientific works include a) the development of a theory relating heat and electricity b) the invention of an electric heater c) the measurement of the force between electric charges d) the invention of the lightning rod.

10. John Milton was very much affected by the science of his time, and *Paradise Lost* shows the influence of the contemporary cosmology. When Milton wrote of the Tuscan artist looking through his glass, the "glass" referred to a telescope. The Tuscan artist was the scientist a) Tycho Brahe b) Evangelista Torricelli c) Galileo Galilei d) Giordano Bruno.

11. Newton's discovery that white light is composed of the

different colors of the spectrum had a great influence on a style of painting known as **a)** fauvism **b)** impressionism **c)** expressionism **d)** chiaroscuro.

12. Sir Arthur Conan Doyle is best known for his authorship of *The Adventures of Sherlock Holmes.* He had another vocation, however, which was **a)** mathematics **b)** medicine **c)** paleontology **d)** drug synthesis.

13. The title of Ray Bradbury's *Fahrenheit 451* refers to the temperature at which **a)** alcohol vaporizes **b)** human beings cannot survive **c)** paper burns **d)** thermometers melt.

14. The title of Josef Strauss's waltz, "Music of the Spheres," refers to **a)** music so heavenly that, listening to it, one is transported to other spheres **b)** ancient cosmology in which the heavenly bodies made music as they traveled in their orbits **c)** the delicate clanging made by little spheres in a kind of scientific pinball game, popular at the time **d)** the sounds made by the spheres in a glass harmonica.

15. A work in which one of the main characters is a computer is **a)** *The Age of Uncertainty* **b)** *2001: A Space Odyssey* **c)** *The End of the Road* **d)** *The Money Game.*

16. The union of art and science in a single work might be epitomized by *Aristotle Contemplating the Bust of Homer.* The work was executed by **a)** Rembrandt **b)** David **c)** Michelangelo **d)** Praxiteles.

17. The modern term for the biblical *brimstone* is **a)** carbon monoxide **b)** calcium carbonate **c)** sulfur **d)** molasses.

18. A work for chamber orchestra that may have been inspired by contemplation of the revolution of the Earth about the sun was written by **a)** Antonio Vivaldi **b)** Arcangelo Corelli **c)** Johann Nepomuk Hummel **d)** Orlando di Lasso.

19. *Density 21.5,* by Edgard Varèse, was composed for **a)** an obese violinist **b)** a thick-headed pianist **c)** a platinum flute **d)** an electronic instrument.

••9••

The Women's Questions

1. In *Silent Spring,* published in 1962, Rachel Carson sounded one of the first public warnings a) about the limits to our fossil fuels b) against certain foods as containing "empty calories" c) against the indiscriminate use of insecticides d) about reducing the decibel levels around the time of the Ides of March.
2. A branch of medicine that deals with the diseases and hygiene of women is a) androgyny b) misanthropy c) gynecology d) ontogeny.
3. Margaret Mead's anthropological work did not take her to a) New Guinea b) Bali c) Samoa d) the Galapagos Islands.
4. Sunday and Monday are named for the sun and moon. Other days of the week are named for Germanic equivalents of gods for whom the planets are named. One day of the week is named for the Germanic equivalent of Venus. That day is a) Tuesday b) Wednesday c) Thursday d) Friday.
5. Jane Goodall has devoted years of patient effort to a) teaching dolphins to communicate with human beings b) observing the behavior of chimpanzees c) studying the stone-age people, the gentle Tasaday d) stalking the yeti.
6. A woman has which of the following pairs of chromosomes? a) XX b) YY c) XY d) YX.
7. The crystallographer who figured prominently in the work that led to the discovery of the structure of the

DNA molecule (it is often stated that she did not receive adequate recognition for that work) was a) Rosalind Franklin b) Oveta Culp Hobby c) Margaret Thatcher d) Jocelyn Bell.

8. Women are rarely afflicted with hemophilia, but can transmit it genetically. Another condition transmitted by women is a) a propensity to baldness b) sensitivity c) obesity d) a widow's peak.
9. Still another characteristic believed to be *sex-linked* is a) visual acuity b) color blindness c) emotional volatility d) intelligence.
10. Chien-Shiung Wu, recent recipient of the Wolf Prize awarded by the State of Israel, is a) head of the World Wildlife Fund b) an experimental nuclear physicist and former president of the American Physical Society c) an ethologist d) a psychologist specializing in emotional differences between Americans and Chinese.
11. Ruth Benedict is best known for her book a) *Patterns of Culture* b) *Patterns of Behavior* c) *Life Is with People* d) *Lives of a Cell.*

12. A recent holder of the position of director of the Royal Greenwich Observatory is **a)** Tuesday Weld **b)** Toni Morrison **c)** Anna Russell **d)** Margaret Burbidge.

13. In France, perhaps because Marie Curie provided a role model, a greater percentage of women appear to enter the physical sciences. Mme. Curie's own daughter Irène was also a Nobel laureate who worked in the field of **a)** artificial radioactivity **b)** quantum electrodynamics **c)** geophysics **d)** plasma physics.

14. As a child psychologist, Melanie Klein **a)** made use of "night music" to soothe disturbed children **b)** developed methods of play therapy in which toys represented father, mother, and siblings **c)** stated that the child is father of the man **d)** had great success with children under the age of two by using what is now known as the Klein bottle.

15. Mary Douglas Leakey, the British archaeologist, is primarily associated with **a)** having dubbed the Piltdown Man a fake **b)** the discovery of the bones of Peking Man **c)** the discovery of the cave at Lascaux **d)** the discovery of the Zinjanthropus skull.

16. Dorothy Crowfoot Hodgkin, British Nobel laureate in 1964, is noted for work in a) astrophysics b) crystallography c) meteorology d) mathematics.
17. Occasionally, although they did not participate in science themselves, women in a position to sponsor science and scientists did so. A ruler associated with the great Swiss mathematician Euler was a) Catherine the Great b) Elizabeth I c) Elizabeth II d) Josephine.
18. The psychoanalyst author of *Our Inner Conflicts* and *The Neurotic Personality of Our Time*, who differed with Freud on some major points such as penis envy, was a) Karen Horney b) Karen Black c) Karen Donner d) Karen Blixen.
19. Anna Freud is noted for a) the Electra complex b) child therapy c) lay analysis d) writing science for the lay reader.
20. Which of the following was not trained in physics? a) Eve Curie b) Maria Goeppert Mayer c) Lise Meitner d) Rosalyn Yalow.
21. Florence Nightingale is known primarily as the heroic nurse of the Crimean War. She is less known for being one of the first to have a) studied the range of the human singing voice b) initiated the training of carrier pigeons for use in wartime communications c) studied the causes of beri-beri d) promoted the use of statistics in medicine and public health.
22. A disease that has long been known to be particularly dangerous to the fetus if contracted by a woman in certain months of her pregnancy is a) salmonella b) rubella c) botulism d) roseola infantum.

• • 10 • •

Going Public

1. Medical officer of the Food and Drug Administration, Frances Kelsey was the recipient of the President's Award for Distinguished Civilian Service in 1962 for a) seeing that cigarette packages bore warnings about the dangers of smoking b) refusing to approve commercial distribution of thalidomide in the U.S. c) warning against the carcinogenic nature of certain hair dyes d) triggering nationwide interest in jogging by warning that Americans were overweight and prone to heart attacks.
2. The position of science adviser to the president was eliminated by President a) Kennedy b) Johnson c) Nixon d) Carter.
3. The director of the Manhattan Project, which produced the first atomic bomb, was a) Judge Irving Kaufman b) Professor Henry Smyth c) General Leslie Groves d) Harvard President James B. Conant.
4. The scientific director of the Manhattan Project was a) Enrico Fermi b) J. Robert Oppenheimer c) Edward Teller d) Stanislaw Ulam.
5. A British writer who has written books on the relationship between science and government is a) C. P. Snow b) E. B. White c) the Earl of Snowden d) Desmond Morris.

6. A government agency which makes decisions on expenditures for science bears the initials a) ABM b) IBM c) OMB d) CBS.
7. The number of physicians per 1,000 population in the U.S. today is *roughly* a) .05 b) 1.5 c) 5 d) 10.
8. The person whose name is associated with bringing about the social and political acceptance of birth control was a) Emma Goldman b) Emma Lazarus c) Flora Gibson d) Margaret Sanger.
9. A government agency which has as one of its concerns the abatement of noise is the a) EPA b) OPA c) IND d) BMT.
10. Brookhaven National Laboratory, on Long Island in New York, is administered by a) a congressional committee b) a board of directors composed of the presidents of several large technical corporations c) a consortium of universities d) nine old men.
11. Project Plowshare dealt with a) the free distribution of modern farm implements to third world nations b) increasing the world food supply by the use of radioactive materials in the soil c) the eradication of diseases such as smallpox by the distribution of vaccines through the World Health Organization d) the peaceful uses of atomic energy such as blasting canal beds with atomic bombs.
12. The conferences in which scientists from all parts of the world get together to discuss matters relating to disarmament bear the name a) Pugwash b) Hogwash c) Pugnose d) Oshkosh.
13. The National Bureau of Standards, which determines the national standards of weights and measures and carries on research toward the improvement of standards and methods of measurement, was established in about the year a) 1800 b) 1850 c) 1900 d) 1950.
14. The amount spent on research and development in this country, as a percentage of the gross national product, is

roughly a) .025% b) .25% c) 2.5% d) 25%.

15. CERN is the acronym for a) the multination nuclear research laboratory located in Switzerland b) the organization of concerned scientists who believe that technology has been permitted to run amuck c) the group of scientists that has been monitoring the freedom of scientific inquiry in different countries d) the Council of European Reactor Nations.

16. Archimedes devoted his considerable talents to the defense of his country. During the war with the Romans, he allegedly a) set the enemy ships on fire by focusing the sun's rays on them with mirrors b) deciphered the enemy code which revealed the Romans' battle plans c) figured out which materials to use for the construction of warships for optimum maneuverability d) constructed underwater craft that could attack enemy ships.

17. The metric system was introduced during the rule of a) Louis the Fat b) Napoleon Bonaparte c) Charles the Simple d) Charles the Bold.

18. The establishment of land-grant colleges, which many believe to be one of the most far-reaching acts of the federal government, originally was a) to offer programs in engineering, agriculture, and home economics b) to teach farmers the elements of mathematics so that they could figure out how much land they had been granted c) to teach surveying d) to utilize for educational purposes land that U. S. Grant had wrested from the South.

19. At the conclusion of the Second World War, American scientists who wanted a forum on science and public affairs founded a journal. The name of that journal is a) *New Scientist* b) *The Physical Review* c) *Daedalus* d) *The Bulletin of the Atomic Scientists.*

20. The cover of the magazine in the previous question carries a picture of a) the peace symbol b) a clock with hands indicating a few minutes before midnight c) weapons being beaten into plowshares **d)** a dove with an olive branch in its beak.

21. A wide-ranging critic of American society, whose targets include the American medical profession and health practices, writes under the name **a)** Ivan the Critical **b)** Ivan the Terrible **c)** Ivan Illich **d)** Ivan Skivitsky Skivar.

••11••

Picture Show

This quiz comes in two parts. First match the pictures with the names listed below; then match the descriptions to the names.

a. *Jonas Salk*
b. *George Washington Carver*
c. *Albert Einstein*
d. *Gregor Mendel*
e. *Jean Piaget*
f. *Benjamin Banneker*
g. *Niels Bohr*
h. *Isaac Newton*
i. *René Descartes*
j. *Andrei Sakharov*
k. *Archimedes*
l. *Nicholas Copernicus*
m. *George Wald*

1. Although this scientist was dedicated to world peace, his work gave rise to a powerful instrument of war. An eclipse of the sun helped verify one of his theories.
2. A U.S. president might not have been crippled had this scientist's work been available earlier. An institute in California bears his name.
3. A monk who could be said to have produced generations of offspring, he used probability theory to formulate the laws of heredity.
4. Famous for his studies of children, this scientist developed a theory about the sequence in which abstract concepts are learned.
5. This mathematician, astronomer, and political philosopher (and author of an almanac much praised by Jefferson) was appointed by Washington to a commission to lay out the new capital.
6. The author of more than one million words on the Bible, once Master of the British Mint and an expert on coun-

terfeiting, this scientist's laws govern our every motion.

7. A renowned institute that bears this physicist's name has been partly supported by beer. The first to explain the energy levels of the hydrogen atom, this Danish scientist was the patriarch of modern quantum theory.
8. A Harvard professor who was an early opponent of the war in Vietnam, this biologist's field is vision.
9. A Polish canon. Others later feared getting "burned" because of his theory.
10. This dissident, the father of the Soviet H-bomb, was awarded the Nobel peace prize by the Norwegian Parliament.

G H I

J K L

M

11. Someone who existed by virtue of his thought. A system of coordinates does him honor.

12. Kids owe this agricultural chemist, who stimulated the economy of the South, half their favorite sandwich.

13. A Sicilian Greek who was one of the first to be engaged in government-sponsored military research. This person's buoyancy proved an asset to the detection of jewel fakes.

••12••
Head Hunting

Here's another quiz in two parts. First match the pictures with the names listed below; then match the descriptions to the names. The answers will reveal who's who.

a. *J. Robert Oppenheimer*
b. *Clara Barton*
c. *Pythagoras*
d. *Galileo Galilei*
e. *Tsung Dao Lee*
f. *Charles Darwin*
g. *Sigmund Freud*
h. *Marie Curie*
i. *Chen Ning Yang*
j. *Paul Ehrlich*
k. *Enrico Fermi*
l. *James D. Watson*

1. This person wrote on Moses and Leonardo, was responsible for an increase in the sale of couches, and calls to mind the song, "Vienna, City of My Dreams."
2. This founder of a cult that worshiped numbers was a religious teacher who studied musical intervals and proved an important theorem in geometry involving squared legs.
3. This professor at the State University of New York at Stony Brook shared a Nobel prize for one of the major advances in modern physics, the statement that parity is not conserved.
4. Free association with this person's name brings forth Paul Muni, a social disease, and 2 times 303.
5. The author of Newton's first law of motion, this scientist measured the height of mountains on the moon, looked at the sun and saw spots before his eyes, and was a noted recanter.

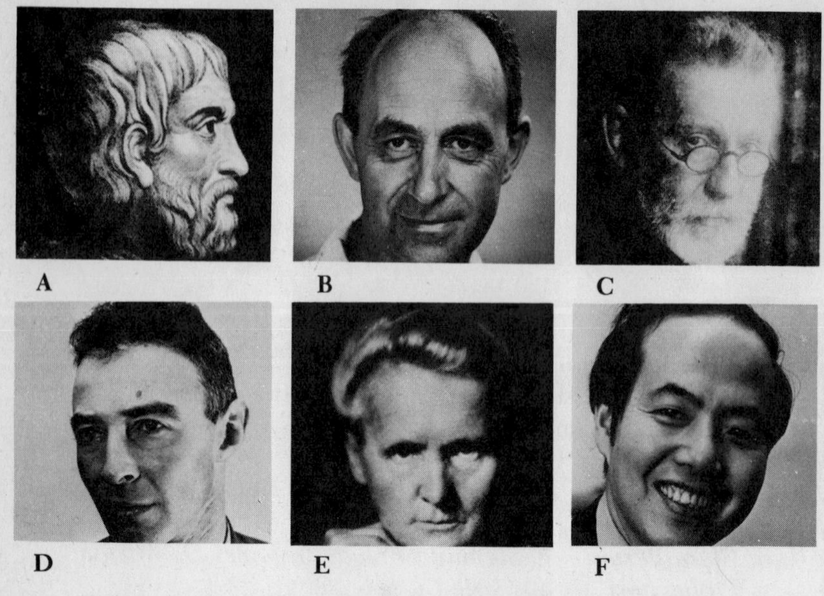

6. This person took a celebrated voyage on a dog of a boat, is linked professionally with Alfred Russel Wallace, and is thought by many to have made monkeys of us all.
7. A unit named for this scientist represents 3.7×10^{10} disintegrations per second. Elements are named after this scientist, the scientist's country of origin, country of residence, and city of residence. This Nobel laureate was also the spouse, parent, and parent-in-law of Nobel laureates.
8. This was the person referred to in the message indicating that the first engineered chain reaction had been achieved: "The Italian navigator has landed in the new world." This scientist was the originator of the theory of beta decay. *Atoms in the Family* was written, as might be expected, by a member of the family.
9. This Civil War "Angel of the Battlefield" worked behind the German lines in the Franco-Prussian War for the International Red Cross. This person organized the

American Red Cross, and pushed for a Red Cross role in catastrophes other than war.

10. This Columbia University professor shared the 1957 Nobel prize in physics for stating, in essence, that nature can distinguish between left-handedness and right-handedness.

11. The biologist half of an interdisciplinary team, this author of a personal account of a twisted molecule unraveled the fabric of life.

12. Familiar with neutron stars and congressional investigators, this person worked in New Mexico on a project named for New York, and quoted, under appropriate circumstances, from the *Bhagavad-Gita:*

> *If the radiance of a thousand suns*
> *Were to burst into the sky*
> *That would be like the splendor of the*
> *Mighty One. . . .*
> *I am become death, the shatterer of worlds.*

••13••

The Art of Science

1. *Nature and Nature's laws lay hid in night: God said, Let Newton Be! and all was light.* These lines **a)** are from the book of Genesis **b)** were written by Alexander Pope **c)** are from Shakespeare's *The Tempest* **d)** were written by John Milton.
2. Galileo has been the source of inspiration for several writers. One who wrote a play with the title *Galileo* is **a)** Maxwell Anderson **b)** Bertolt Brecht **c)** Arthur Miller **d)** Jean-Paul Sartre.
3. A group portrait known as *The Anatomy Lesson of Dr. Tulp* was painted by **a)** Leonardo **b)** Vermeer **c)** Rembrandt **d)** Cézanne.
4. Ben Johnson wrote the play *The Alchemist* in which he satirizes a group prominent in his time. A major objective of the alchemists was **a)** to discover a chemical that would assure eternal youth **b)** to produce gold from the base metals **c)** to deduce the laws of chemistry from the laws of physics **d)** to produce potions that would enable people to experience altered states of consciousness.
5. Whom would Walt Whitman not have used as the model for his "When I Heard the Learn'd Astronomer"? **a)** Herschel **b)** Halley **c)** Hubble **d)** Harvey
6. The painter who employed Newton's discovery that

white light consists of all the different colors of the spectrum, putting dots of color in juxtaposition and letting the eye combine them into the desired color—a technique called pointillism—was **a)** Matisse **b)** Picasso **c)** Seurat **d)** Van Gogh.

7. The author of the *Novum Organum* is known as the father of the scientific method, although he did only one experiment and that one killed him. (He caught cold while stuffing a fowl with snow to observe the effect of chilling on the preservation of the flesh.) He was **a)** Francis Bacon **b)** Roger Bacon **c)** Francis Drake **d)** Roger Baldwin.

8. The title of a novel by Thomas Pynchon could bring to mind a song made famous by Judy Garland, and the scientist **a)** Michael Faraday **b)** Dmitri Ivanovich Mendeleev **c)** Charles Darwin **d)** Isaac Newton.

9. The composer of such scientifically titled pieces as *Métal, Ionisation, Hyperprism,* and *Intégrales* is **a)** Alban Berg **b)** Olivier Messiaen **c)** Francis Poulenc **d)** Edgard Varèse.

10. What character in literature actually existed and wrote an early piece of science fiction involving a trip to the moon? **a)** Cyrano de Bergerac **b)** Falstaff **c)** Captain Nemo **d)** Scrooge.

11. The nuclear-powered submarine *Nautilus* was named after **a)** an undersea craft by the same name in a Jules Verne work **b)** a figure in Norse mythology **c)** a poem by Keats written about a shell **d)** a character in a folk tale who stayed under water longer than anyone else.

12. An orchestral suite inspired by the solar system was written by **a)** Dmitri Shostakovich **b)** Arnold Schoenberg **c)** Sergei Prokofiev **d)** Gustav Holst.

13. What twentieth-century writer supported himself, before he became well known, by teaching mathematics and physics? This background shows in one of his works in which the protagonist is a mathematician. **a)** Albert Camus **b)** Thomas Mann **c)** Vladimir Nabokov **d)** Alexander Solzhenitsyn.

14. John Keats, in his poem, "Lamia," complains that a natural phenomenon has been spoiled for him because he has learned how it is produced. That natural phenomenon is **a)** a cloud **b)** lightning **c)** a rainbow **d)** the northern lights.

15. The person who was both artist and scientist was **a)** Leonardo **b)** Rembrandt **c)** Gauguin **d)** Delacroix.

16. Jules Verne wrote about rockets in space and used as his launching site one that is actually used today. That is **a)** Los Alamos, New Mexico **b)** Cape Canaveral, Florida **c)** Houston, Texas **d)** White Sands, New Mexico.

17. A modern piece of piano music that might be suggested by the Big Bang theory, a microsecond after the big bang, was written by **a)** Béla Bartók **b)** Benjamin Britten **c)** Bohuslav Martinu **d)** Aram Khachaturian.

18. The doctor in the novel by Mary Wollstonecraft Shelley was named **a)** Jekyll **b)** Hyde **c)** Frankenstein **d)** Lugosi.

19. In which Sinclair Lewis novel is the hero a doctor seeking to cure bubonic plague in the West Indies? **a)** *Main Street* **b)** *Dodsworth* **c)** *Babbitt* **d)** *Arrowsmith.*

20. The discovery of the structure of DNA is the subject of the book **a)** *The Structure of Scientific Revolutions* **b)** *Revolutions in Science* **c)** *The Double Helix* **d)** *The Spiral Staircase.*

21. A book written by the medical doctor Lewis Thomas which earned the 1974 National Book Award for Arts and Letters has the title **a)** *The World Within* **b)** *The Lives of a Cell* **c)** *The Sex Life of an Aphid* **d)** *The Life and Times of a Paramecium.*

22. A very recent opera has the title **a)** *Einstein on the Beach* **b)** *Newton and Milton at the Royal Society* **c)** *Galileo at the Inquisition* **d)** *Kepler at the Ball.*

23. The great Russian writer Anton Chekhov was a practicing physician. Name three other well-known doctor-writers.

••14••
Frags

In preparing quizzes, we are often led into byways fascinating to us but which, because they are too specialized or not scientific, we do not use. In producing "Frags" (fragments of scientific words or phrases which the reader was to complete), we got caught up in fragments with three and four identical letters in a row. Some appear here; others, like PP P in umklaPP Process, were too specialized, and LaPP Pedants and LeTT Teachers are not necessarily related to science.

The few fours-in-a-row we got were in two words; they included a smaLL LLama and thrEE EEls. Then along came a postcard with "an exhaustive compilation" of "perfectly ordinary English words having four identical letters in a row." (1) buLLLLama—the male of a beast of burden common to the Peruvian Andes; (2) trEEEEl—a small elongated arboreal snakelike salamander (Louisiana); (3) zOOOOze—a primordial slime which served as a breeding place for primitive animal organisms in the Precambrian epoch. Our correspondent, who signed his name Herrmann Massstab, stated, "No further words of this type are possible in English. Don't bother to try."

Well, our search thus far has failed to locate any of these creatures (or Mr. Massstab, for that matter), but we feel they sound so reasonable that if they don't exist, they should. Anyway, here is a more pedestrian set of fragments. The letters in each fragment appear in the sequence in which they occur in the complete words, and they are consecutive. Spaces indicate breaks between words, not deleted letters. (This quiz may be harder than most, so don't be discouraged over a flagging score.)

1. EE E Two words from thermodynamics that make one think of some of the best things in life and thumbing one's nose at the utilities companies. (For some, the latter action may be included in the former.)
2. GHT P Two words that come up in fiber optics and could be used by feeble smokers.
3. THM Type of scale often used for graphs (nonlinear), which compresses data. If applied to a dancer, he or she would be a klutz.
4. LL L Two words of interest to geologists and skiers. Also, garment industry jargon for autumn fashions.
5. C PR In thermodynamics, these two words denote a change in state of a system (normally a gas), which occurs without heat entering or leaving the system. Sounds like a helpful procedure.
6. SCHR The name of one of the inventors of quantum mechanics and perhaps the most important equation in modern physics. It sounds like the name of the character in *Peanuts* who plays Beethoven.

7. ST M Two words in the theory of relativity for a characteristic of a body when the body is not moving (in the simplest interpretation). In everyday life, one usually amasses too much of the second if one does too much of the first.
8. Q V This measure of frequency selectivity consists of two terms that might be applied to the worth of the stick used in billiards.
9. LL L The same fragment as in item four here denotes a rare phenomenon of atmospheric electricity. If there were such a thing as a spherical thunderstorm, it might exhibit this. One of the 1978 Nobel laureates in physics, Pyotr Kapitsa, did work on this subject, which has been offered as an explanation of some UFOs.
10. BSTR This summary or abridgment is found at the beginning of many scientific publications. The word also describes some art and thought.
11. FT L A controlled operation in space science, arbitrarily defined as one in which the approach velocity is less than 50 feet per second. In more down-to-earth situations, these two words might be used to describe no bounce.

12. THR Used to denote the beginning or entrance to all kinds of things from frequencies to homes. In the latter, it figures in the transportation of brides.
13. CK W Two words for a pressure disturbance that propagates in a medium such as air and can sometimes cause windows to break, or an indication that astonishment abounds.
14. RPT A word that appears in a number of fields, optics and acoustics among them. Absent-minded professors supposedly commit their follies because of this, and paper towelings their triumphs.
15. G B Two words for a theory of the origin of the universe. And God said, "Don't whimper!"
16. ND W The broadness of a portion of the electromagnetic spectrum. These two words would also be of interest to someone photographing a group of musicians and having difficulty fitting them all into a single picture.
17. P SP Two words one might hear in connection with the Voyager missions that mean *far out* but would seem to mean *far in.* They might also describe a crevasse.
18. TCH Term used in music and baseball. In the former, it is related primarily to the frequency of vibration.
19. DSP An electroacoustic transducer intended to radiate acoustic power into the air from an electrical input, it is part of a piece of equipment commonly found in homes. It is also the opposite of a whisperer.
20. ALL ALL ALL One of our detours mentioned earlier took us to the construction of repeating *sets* of letters. (For example, ELELE L could be ukELELE Lady.) Here three sets are contained in three words that might make up a message sent by a metallurgical laboratory aping the procedure of automobile companies that discover they've sold cars with faulty parts. The last word is a mixture of metals.

••15••

The Old Math

1. In the United States, one billion is 10^9. In England, one billion is 10^{12}. Apart from the shock given an unknowing Britisher reading of our national debt, there was considerable confusion in high-energy physics, where energies are often measured in units of 10^9 electron volts. To eliminate the scientific confusion, at least, the unit GeV (G for giga) was adopted to represent 10^9 electron volts.

   In the U.S., the number system is based on groups of three zeros added to the three zeros in 1,000. One *bi*llion is two sets of three zeros added to the three zeros in 1,000—that is, 10^9—while one *tri*llion is three sets of three zeros added to the three zeros in 1,000—that is, 10^{12}. The American centillion—when used more precisely than merely to indicate a very large number—would then be **a)** 10^{100} **b)** 10^{103} **c)** 10^{300} **d)** 10^{303}.

2. The expression HEFG, compared to EFGH, is called a

a) permutation b) combination c) perturbation d) consternation.

3. 10^{100} is known as a a) barney b) googol c) yankee d) doodle.
4. A hexagon a) has ten sides b) was a geometric figure employed by witches to surround a person on whom a spell was being cast c) has eight sides and eight angles d) has six sides.
5. The number $3^{1/2}$ is a) rational b) irrational c) erratic d) quadratic.
6. If one of the angles in a right triangle is θ, sin θ is a) the ratio of the side opposite θ to the hypotenuse b) the ratio of the side adjacent to θ to the hypotenuse c) the ratio of the side opposite θ to the side adjacent to θ d) $+1$ if θ is positive and -1 if θ is negative.
7. A 3,4,5 triangle a) is one with angles in the ratio of 3 to 4 to 5 b) is an isosceles triangle c) is a right triangle d) does not exist.
8. A polygon with five sides and five angles is a a) pentagon b) pentathlon c) pentahedron d) pentateuch.
9. Computer languages do not include a) Basic b) Fortran c) Algol d) Logla.
10. The story is told about Gauss that as a small child he was assigned the task of finding the sum of all the integers from 1 to 100. There proved to be little labor in it for the miniature mathematician; he mentally regrouped the integers into $1 + 100, 2 + 99, 3 + 98 \ldots$ and wrote down the sum a) 4,950 b) 5,000 c) 5,050 d) 5,500.
11. A basic topic treated by integral calculus is a) whole functions b) the area under a curve c) integrated circuits d) holistic relations.
12. A primary topic in differential calculus is a) how to distinguish between different types of functions b) discriminants c) the slope of a curve at a particular point d) wage increments.

13. A phrase that might arise in a conversation between professional mathematicians is a) squaring the circle b) Circle in the Square c) cubing the square d) curbing the dog.
14. nx^{n-1} is a) the derivative of x^n b) the integral of x^n c) the PaLace transform of x^n d) the nth root of x^n.
15. Geometry supposedly arose a) from astronomical studies by the Phoenicians b) in connection with studies of geological formations c) from measurements of land areas by the ancient Egyptians d) in studies of buoyant forces by Pythagoras's rival, Geometricus.
16. $4/3\ \pi r^3$ is the volume of a) a pie-shaped segment of a disk of radius r b) a cylinder of radius r and height π c) a sphere of radius r d) an ellipsoid of revolution of major axis r and minor axis π.

Nurit

••16••

"A Jug of Wine, A Loaf of Bread— And Thou Beside Me Singing . . ."

1. The science that deals with wine and wine-making is **a)** etiology **b)** ethology **c)** oenology **d)** rheology.
2. The length of time it takes to roast a piece of beef in the oven, at a particular temperature, **a)** depends only upon the weight of the meat **b)** depends only upon the shape of the meat **c)** depends upon the weight and shape of the meat **d)** depends upon whether the oven operates on electricity or gas.
3. A popular classic that strongly suggests the Earth's rotation on its own axis was written by the composer **a)** Victor Herbert **b)** George M. Cohan **c)** Irving Berlin **d)** Cole Porter.
4. A scientist associated with research on the fermentation of wine was **a)** Louis Pasteur **b)** James A. Bacchus **c)** Robert Koch **d)** Dyan Isis.
5. An astronaut in a gravity-free environment wishing to drink some juice **a)** would turn a kind of somersault, in order to be atop the container of juice **b)** would turn the container upside down and drop the contents into his mouth **c)** would use a straw **d)** could drink in the normal terrestrial manner, from a glass.

6. Occultation by the moon is used by astronomers as a precise method of locating astronomical objects. Occultation *of* the moon might be suggested by a song made popular by **a)** Dinah Shore **b)** Kate Smith **c)** Peggy Lee **d)** Doris Day.
7. The statement *100 proof* on a bottle of liquor means that the contents **a)** are 100% natural ingredients and no preservatives **b)** are 100% alcohol **c)** have been certified by a government agency to contain no wood alcohol **d)** are 50% alcohol.
8. One of the early applications of computers was to language translation. The story is told that the phrase, "The spirit is willing but the flesh is weak" was given to a computer for translation into Russian. When translated back into English, what emerged was, "The vodka is good but the meat is so-so." The word *vodka* comes from the Russian word *voda,* which means **a)** earth **b)** air **c)** fire **d)** water.
9. Pots with copper bottoms are employed in cooking because **a)** copper, a good conductor, distributes the heat and prevents scorching **b)** copper has a high specific heat and holds heat better than other materials **c)** these pots simply caught the imagination of the public and became a status symbol **d)** copper best reflects the light from the flame.
10. A fortified wine **a)** has alcohol added to it in addition to that produced by fermentation **b)** is so named because, being stronger than ordinary wines, it fortifies you for a journey **c)** is one that remains in the cellars or "fortifications" longer than do ordinary wines **d)** contains methyl alcohol.
11. Wines contain significant amounts of **a)** H_2O_2 **b)** C_2H_5OH **c)** H_2SO_4 **d)** CH_3COOH.
12. A popular song of several decades ago that dealt with light of about 4,700 angstrom units had the title **a)** "Red Sails in the Sunset" **b)** "She Wore a Yellow Rib-

bon" c) "Blue Moon" d) "Jeannie with the Light Brown Hair."

13. A 12-ounce can of beer a) contains about as much alcohol as a one-ounce shot of 100-proof whiskey b) contains less than half the alcohol in a shot of whiskey c) contains twice as much alcohol as does a shot of whiskey d) contains a negligible amount of alcohol compared to a shot of whiskey.
14. Wine-making a) can be traced back to the earliest civilizations b) appeared in Europe in the Middle Ages c) was not developed until the subject of chemistry was understood d) arose with Christianity in connection with the sacrament.
15. Often after eating pizza, the front of one's palate is in shreds. One reason is: a) The pizza is highly spiced. b) The pan the pizza is on retains the heat. c) The high specific heat of the ingredients tends to hold the heat. d) The high density of the ingredients causes them to conduct the heat directly to your palate.
16. A cocktail wine, such as a sherry, normally contains approximately what percentage of alcohol? a) 5% b) 10% c) 20% d) 40%.
17. A song that may suggest the vernal equinox in the western part of the United States is concerned with a) Yellowstone National Park b) the Rocky Mountains c) Lake Tahoe d) the Painted Desert.
18. A table wine normally contains about what percentage of alcohol? a) 3.2% b) 10% c) 25% d) 40%.
19. Approximately how many gallons of wine will the average French person drink in his or her lifetime? a) 50 b) 500 c) 5,000 d) 50,000.
20. Mead is made from a) milk b) honey c) molasses d) resin.
21. A song title which advocates anticipation that the Earth's atmosphere will allow increased transmission of electro-

magnetic radiation from roughly 4,000 to 8,000 angstrom units concerns a woman by the name of **a)** Susie **b)** Rosy **c)** Nellie **d)** Jeannie.

22. A drunkometer is **a)** a question and answer test used by psychologists to determine whether a person has tendencies toward alcoholism **b)** a device used by policemen to assess alcohol levels in the blood, to test whether there has been some "spirited" driving **c)** an instrument that measures the needs of alcoholics undergoing detoxification **d)** an instrument used to test the capacity of entrants in beer drinking contests, to avoid fatalities.

••17••
Sound and Music

1. If an angel were to play the violin in the void of outer space, the result would be **a)** a much louder sound than would be heard on Earth **b)** a sound about the same as would be heard on Earth **c)** no sound at all **d)** a heavenly sound.
2. If a person is deaf in one ear, that person **a)** gets only half the pleasure others do in listening to music **b)** is insensitive to loudness **c)** will usually find it harder to detect the direction of a source of sound **d)** cannot hear certain frequencies.
3. Sound intensity is measured in **a)** angstroms **b)** decibels **c)** ohms **d)** parsecs.
4. Sound intensity determines **a)** frequency **b)** loudness **c)** pitch **d)** quality.
5. An organization devoted to making instruments that are the equal of famous old instruments is **a)** The Catgut

Acoustical Society b) No Strings Attached c) String Along d) Strads and Strings.

6. In a trumpet, the sound is initiated by the vibration of a) a reed b) the player's lips c) the player's larynx d) the player's pharynx.
7. The relationship between frequency and wavelength is a) frequency × wavelength = velocity of sound b) frequency/wavelength = velocity of sound c) frequency × wavelength = intensity d) frequency/wavelength = intensity.
8. Which notes were consonant with others was worked out by a) Pythagoras b) Palestrina c) Orpheus d) David.
9. Dogs respond to dog whistles while humans do not a) because the whistles emit mating calls b) because dogs can hear the high frequencies emitted by the whistles while humans cannot c) primarily because dogs have very sensitive ears and can hear sounds much fainter than humans can hear d) because the whistles emit frequencies below 60 cycles per second, which sound like growls.
10. Clamping a violin string at its midpoint before bowing results in a) its fundamental frequency going down an octave b) its fundamental frequency going up an octave c) its fundamental frequency remaining the same d) no sound at all.
11. Music of the spheres a) was believed at one time to be the music played by angels in Heaven b) was the sound made by the heavenly bodies in their orbits, according to the ancient Greeks c) was earthly music, in contrast to the ethereal music heard in Plato's ideal world d) was the music emitted by an ancient instrument consisting essentially of jangling balls.
12. The longest pipes of an organ are the source of a) the organ's highest notes b) the organ's lowest notes c)

notes in the middle range d) no sound at all; they are purely decorative.

13. A piano tuner, in tuning a piano, adjusts a) the tension of the strings b) the length of the strings c) the points where the strings are struck d) the force with which the strings are plucked.
14. The fundamental is a) the highest frequency with which a string can vibrate b) the lowest frequency with which a string can vibrate c) the loudest note a string can emit d) the note most frequently played in a composition.
15. If a tenor were to inhale some helium before he started to sing, his voice would sound a) more like that of a bass b) unchanged c) more like that of a soprano d) breathless.
16. Voiceprints a) are considerably inferior to fingerprints in that they can never make unique identifications b) were discussed in Solzhenitsyn's *The First Circle* c) have been accepted as evidence of identification in law courts for the last fifty years d) have been used by rock singers to copyright their records.
17. The voiceprint method of identification consists of a) measuring the range of intensity in a person's speech in a given time interval b) measuring the wavelengths over which a person's speech ranges c) breaking the voice down into its component frequencies and determining the intensity at each frequency d) having the voice, as an input, print out the name of the speaker.
18. Pitch is related to a) frequency of vibration b) velocity of sound c) overtones d) undertones.
19. A characteristic reverberation time in a concert hall is roughly a) .02 second b) .2 second c) 2 seconds d) 20 seconds.
20. Sound and light are both waves but sound waves are longitudinal while light waves are transverse. That

means **a)** merely that sound waves are longer than light waves **b)** that the vibrations giving rise to sound are perpendicular to the direction in which the sound wave travels while the light vibrations are in the same direction as the light wave travels **c)** that the vibrations giving rise to sound are in the direction the sound wave travels while the light vibrations are perpendicular to the direction in which the light wave travels **d)** that sound waves travel in the direction of the ether while light waves travel across it.

21. Johann Strauss dedicated his "Paroxysmen Waltz" to **a)** doctors **b)** an epileptic friend **c)** medical students **d)** St. Vitus.

22. A person who can identify an isolated tone without reference to any other tones has **a)** absolute pitch **b)** perfect hearing **c)** sound memory **d)** notal recall.

••18••

Famous Formulas

Another matching quiz. First match each expression with a "name" (or associated term) listed below; then match the descriptions to the paired names and expressions. (Note that there are fourteen names and only twelve expressions. Two names are extraneous.) We have used what are probably the most common symbols in the expressions; symbols can vary from text to text. We leave it to the experts to determine when the author of a scientific expression gets to "possess" his expression—as in Coulomb's law—or when the author's name is applied in a more monumental fashion—as in the Schrödinger equation. It isn't the difference between laws and equations, because there are lots of "possessed" equations—Bernoulli's, Euler's—and we doubt that it lies in the author's having only one to his credit, for many were named during the author's lifetime—some even when the author was young—so how would anyone know he wouldn't produce another?

a. *Equation of a straight line*
b. *Law of gravitation*
c. *Lens equation*
d. *The Schrödinger equation*
e. *Coulomb's law*
f. *Grimm's law*
g. *A fission reaction*
h. *A kinematic equation*
i. *Radioactive decay law*
j. *Equation of a parabola*
k. *Equation of an ellipse*
l. *A fusion reaction*
m. *A faction reaction*
n. *Potential energy*

(A) $F = \frac{kq_1q_2}{r^2}$	(G) $-\frac{\hbar^2}{2m}\frac{\partial^2\Psi}{\partial x^2} + V\Psi = i\hbar\frac{\partial\Psi}{\partial t}$
(B) n+U→Sr+Xe+n's	(H) gh, dh, bh→g, d, b g, d, b→k, t, p k, t, p→h, th, f
(C) $N(t) = N_0e^{-\lambda t}$	(I) $y = mx + b$
(D) $V = mgh$	(J) $D + T \rightarrow He^4 + n$
(E) $\frac{1}{p} + \frac{1}{q} = \frac{1}{f}$	(K) $s = \frac{1}{2}at^2$
(F) $F = \frac{Gm_1m_2}{r^2}$	(L) $\frac{x^2}{a^2} + \frac{y^2}{b^2} = 1$

1. It's what "makes the world go round." Though it isn't love, it *is* an expression of the attraction between two bodies. The constant in the expression was measured by a wealthy English eccentric of noble ancestry after whom a renowned laboratory is named.
2. Many a person bears the weight of its results on the bridge of his or her nose. Useful to optical instrument makers, it relates object and image distances.
3. It is what one would use to estimate how long it would take to hit the water after jumping off the Brooklyn Bridge. It was discovered by a person whose first and last names are almost identical.
4. Published in 1926, it is the *sine qua non* in the analysis of atoms; it can be considered the fundamental equation of nonrelativistic quantum mechanics.

5. This is a reaction in which the energy produced per unit mass is millions of times greater than that in a chemical reaction such as the burning of coal. John Hersey wrote a book about one of its effects.
6. Found by a French military engineer who got a charge out of it, it gives the force between two very small spheres, neither electrically neutral, a specified distance apart.
7. A geodesic on a plane, it is also what Lyndon Johnson said Gerald Ford couldn't walk while chewing gum.
8. A general class of which a circle is a special member, it gives the orbit of a planet.
9. What might be termed a bright hope for the future, at present it warms our hearts (and other parts). Studied in a Tokamak, it is one of the reactions in the p-p chain which occurs in the sun.
10. A principle of relationships in Indo-European languages, it was formulated by a founder of comparative philology better known to the public for the fairy tales he collected with his brother.
11. Associated with position, and normally convertible into another form, it is present in a flowerpot perched on a windowsill.
12. An expression of the relations between mothers and daughters, it is useful in areas in which history is irrelevant. Mme. Curie and her daughter Irène Joliot-Curie played a role in its development.

••19••

Chemical Solutions

1. Transuranic elements are a) elements found beyond the Ural Mountains b) ores containing uranium that are imported to this country by sea c) elements with a nuclear charge greater than 92 d) elements found on the planets which orbit beyond Uranus.
2. H_2SO_4 is a) an acid b) a base c) an isomer d) a polymer.
3. A molecule is defined as a) a particular configuration of two atoms b) the center of an atom c) an antibiotic extracted from moles d) an association of two or more atoms.
4. For a given quantity of an ideal gas, Boyle's law a) determines the boiling point; that is, the temperature at which the liquid becomes a gas b) states that the temperature times the volume is constant at constant pressure c) states that the pressure times the volume is constant at constant temperature d) states that the pressure times the temperature is constant at constant volume.
5. The alkali metals include a) sodium b) palladium c) nickel d) aluminum.
6. The law of mass action a) relates the masses on one side of a chemical equation to the masses on the other side at equilibrium b) is a law that deals with large masses of material and holds only approximately for small quantities c) is Newton's second law as applied to chemical reactions d) is a law formulated by Marx and Engels that has been applied to chemistry.

7. Who did *not* win the Nobel prize in chemistry? a) Ernest Rutherford b) Marie Curie c) Amadeo Avogadro d) Svante Arrhenius.
8. Most of the elements that lack one electron that would give them a closed shell electron configuration are called a) anions b) cations c) halcyons d) halogens.
9. Sodium chloride is a) a salt b) a pepper c) an alka seltzer d) a bromide.
10. Oxygen in the air at sea level is generally found in the form of a) atoms b) molecules c) O^{--} ions d) $(OH)^{-}$ ions.
11. The discovery of oxygen is credited to a) Sexton b) Priestley c) Canon d) Bishop.
12. An ion a) is part of an iron compound b) is always negatively charged c) can be either negatively or positively charged d) derives its name from its shape, which resembles a Greek column.
13. Adding salt a) raises the melting point of ice b) lowers the freezing point of water c) lowers the boiling point of water d) reduces the density of water.
14. The number of calories required to convert one gram of water at 100°C into steam is about a) .1 b) 1 c) 5 d) 500.
15. The noble gases are so called because a) they are very rare and refined b) they often emanate from the noble metals c) they were discovered by Alfred Noble d) at the time of their discovery they refused to mix with the common herd of elements to form compounds.
16. The most abundant of the elements in the Earth's crust is a) carbon b) hydrogen c) nitrogen d) oxygen.
17. A base a) has a low density b) has a high pH in solution c) is low on the electromotive series d) has an unpleasant taste in solution.
18. Organic chemistry a) is chemistry that is dynamic and in the forefront of research b) is the chemistry of the

processes in the different organs of the body **c)** deals with carbon compounds **d)** involves the chemical metallurgy of organ pipes.

19. An electrolyte is **a)** a rechargeable flashlight **b)** a nonmetallic conductor of electricity in which the current is carried by the movement of ions **c)** a substance, normally a liquid, which is easily electrified **d)** is an especially light electron.

20. Electrolysis does *not* encompass **a)** plating metals **b)** hair removal **c)** separating water into hydrogen and oxygen **d)** monitoring heart action.

21. The temperature at which the Celsius and the Fahrenheit scales register the same number is **a)** $-273°$ **b)** $-40°$ **c)** $0°$ **d)** $100°$.

22. NH_3 is the chemical formula for **a)** ammonia **b)** chlorine bleach **c)** natural gas **d)** the gas associated with rotten eggs.

23. Which is chemical terminology? **a)** free radical **b)** chained radical **c)** conservative element **d)** moderate solution.

24. An isothermal process **a)** is one that is conducted at very low temperatures **b)** is one that is conducted at a constant temperature **c)** is one in which heat neither enters nor leaves the system **d)** is one in which underwear is chemically treated to make it temperature resistant.

25. A significant source of methane in the atmosphere is **a)** waste products from nuclear power plants **b)** byproducts from sugar refineries **c)** bovine flatus **d)** canine urine.

20

Famous Figures

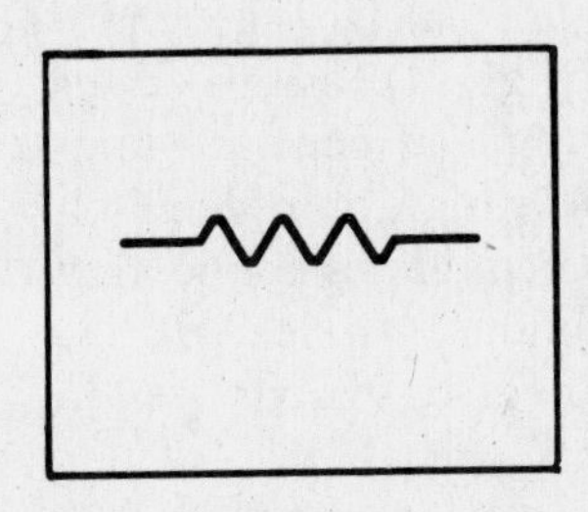

1. This figure is **a)** the electrical symbol of inductance **b)** the mathematical indication of the longest distance between two points **c)** the electrical symbol of resistance **d)** the biological representation of a class of worms of which the earthworm is a member.

2. This figure shows **a)** a prism **b)** a rectangular pyramid **c)** the unit cell of a honeycomb **d)** a pre-Columbian tepee.

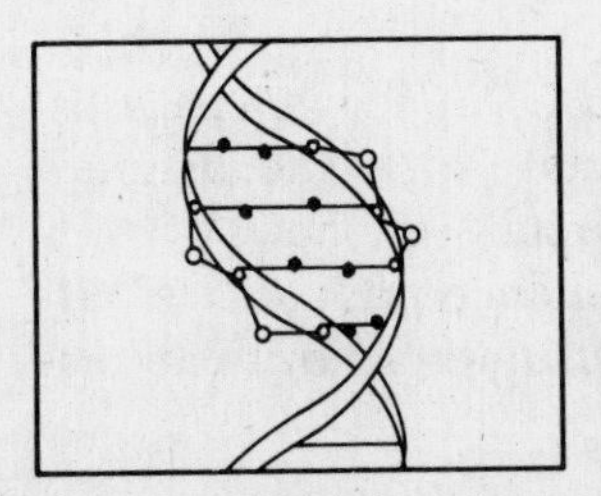

3. This figure shows **a)** the structure of a human hair **b)** the configuration a flexible strip (such as a rope ladder) would take under the action of an asymmetric torque **c)** a Möbius strip **d)** the structure of deoxyribonucleic acid.

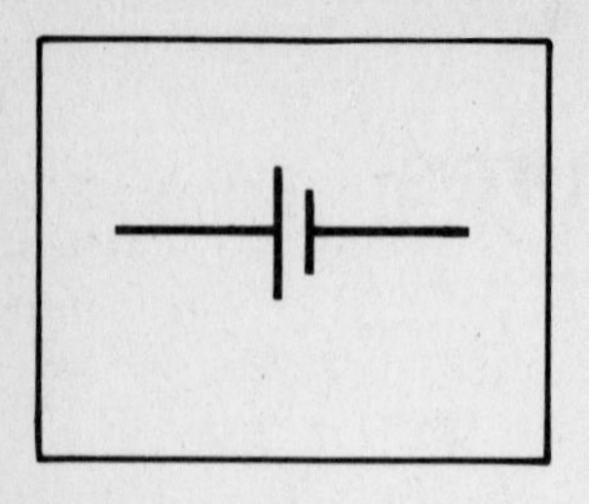

4. This is **a)** the electrical symbol for capacitance **b)** the electrical symbol for a battery **c)** the cuneiform representation of infinity **d)** an illustration of an optical illusion in that you think the right side of the figure is shorter because of its placement on this page.

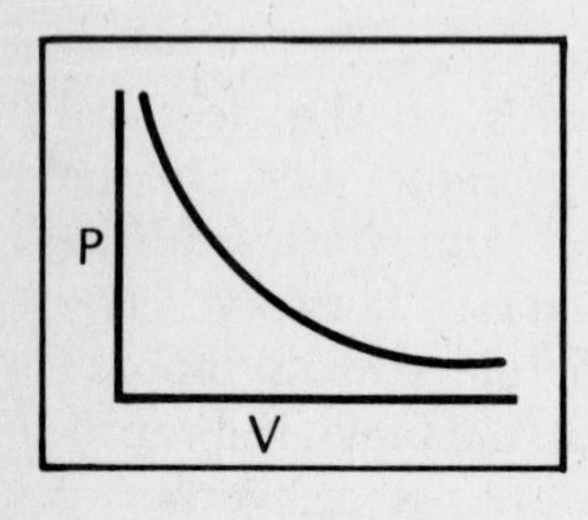

5. This figure **a)** is a graph of Boil's law **b)** gives pressure versus volume for a gas at constant temperature **c)** is a graph of Samuelson's law in economics connecting price and volume of sales **d)** is a graph of pull versus viscous drag in a liquid.

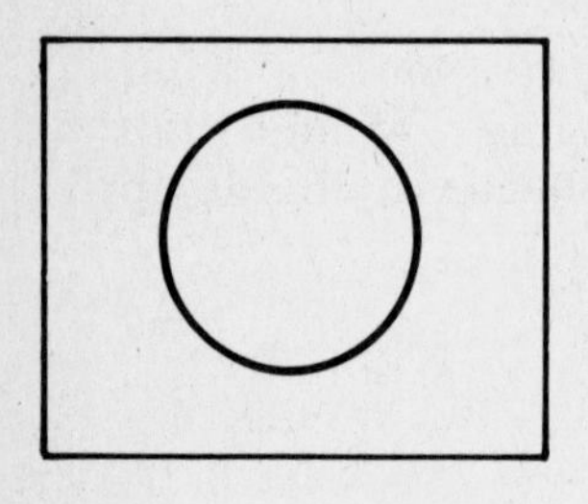

6. Which statement does *not* apply to this figure, a circle? **a)** It represents the maximum area for a given perimeter. **b)** It is a curve having constant radius of curvature. **c)** It is the locus of all points equidistant from the center. **d)** For a radius R the area is $4\pi R^2$.

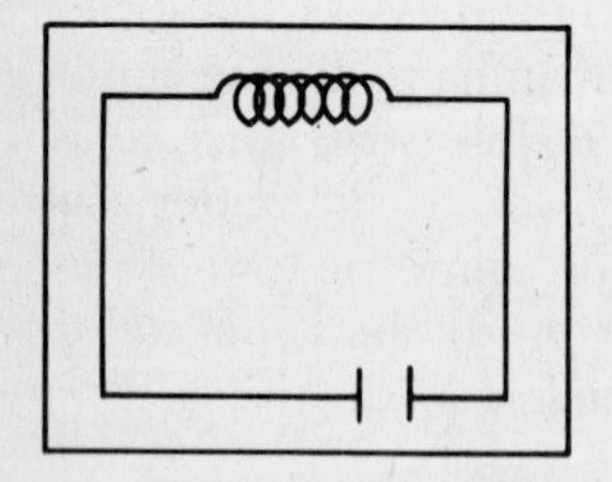

7. This figure **a)** shows a basic oscillator circuit **b)** is the diagram of a short circuit **c)** shows a resistance capacitance inductance circuit **d)** illustrates a Feynman diagram.

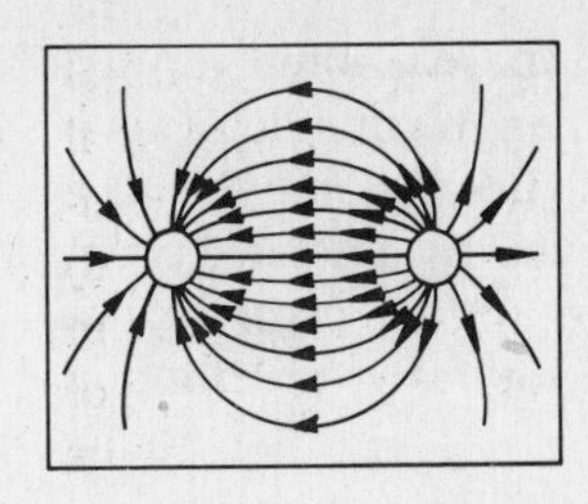

8. This figure **a)** illustrates the pinch effect in a plasma **b)** is a sketch of a boll weevil **c)** shows the electric field due to two equal but opposite point charges **d)** shows the magnetic field in a Tokamak.

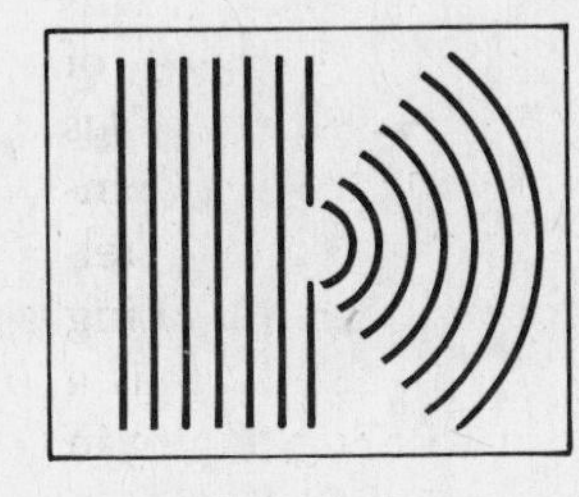

9. This figure **a)** is used widely in experiments in psychology in connection with optical illusions **b)** is part of an ophthalmological test for astigmatism **c)** illustrates Huygens's principle that every point on a wave front acts as a source of secondary wavelets **c)** illustrates a double slit diffraction pattern.

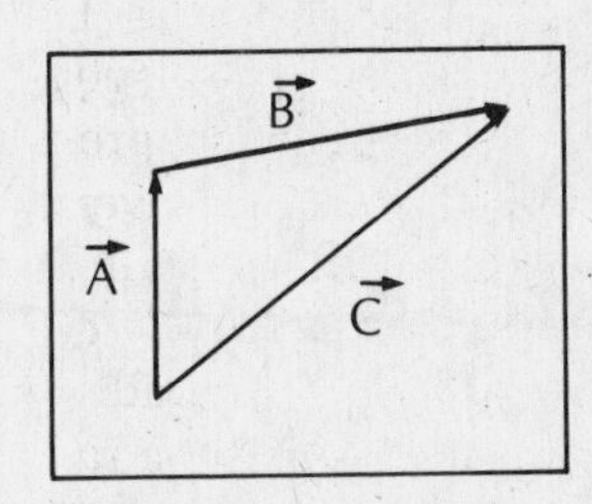

10. This figure illustrates **a)** vector addition **b)** tensor subtraction **c)** vector multiplication **d)** the arrows of time.

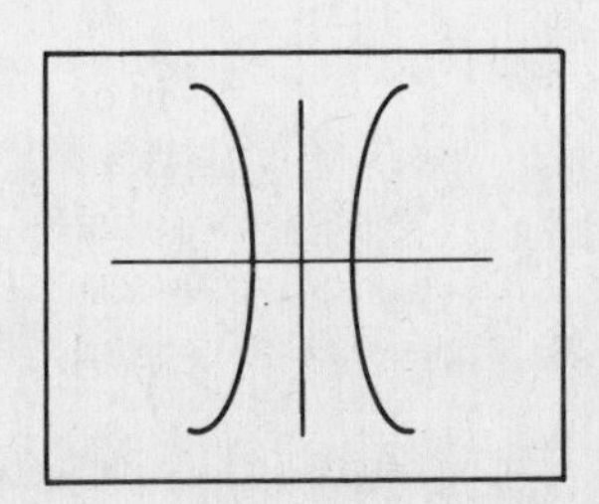

11. This figure is **a)** a parabola **b)** a hyperbola **c)** a symbol used in logic to indicate a gross exaggeration **d)** a graph of the tangent function.

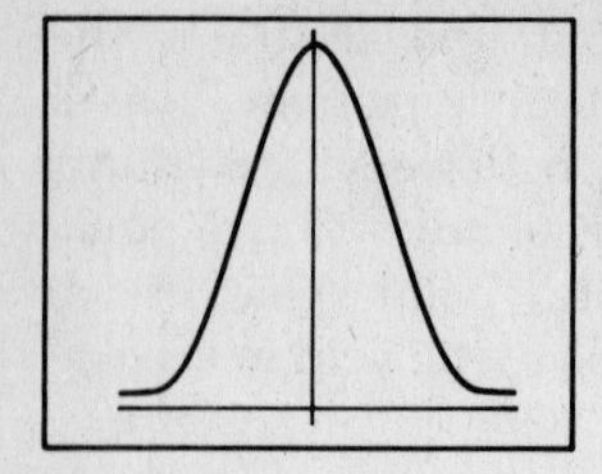

12. Although there is some variety, or looseness, in terminology, one thing this figure is *not* is **a)** a Gaussian curve **b)** a normal probability curve **c)** an error curve **d)** a Poisson, or fishy, distribution curve.

••21••

Homage to Albert

1. In the well-known expression $E=mc^2$, c a) denotes energy b) is the gravitational constant c) is the velocity of light d) indicates that the energy is of low grade.
2. Numerous myths have been perpetrated about Einstein and his schooling—some of them by anxious parents with offspring having difficulties with their studies. One *non*-myth is that Einstein a) was terrible in math b) failed German c) was a high school dropout d) flunked gym.
3. The factor $\sqrt{1-v^2/c^2}$, which appears so often in the special theory of relativity, a) is called a mean square root b) arises in connection with length contraction c) connects the intensity of light with the energy of a photon d) adds great mathematical complexity and is therefore called the root of evil.
4. The character in the comic strip *Alley Oop* that was patterned after Einstein was a) Dinny b) Dr. Wonmug c) Professor Moriarty d) Speedy Alberto.
5. The zigzag movement of small particles suspended in a fluid, first observed by a botanist, was treated by Einstein in his theory of a) Brownian motion b) Jonesian movement c) particular walks d) Brahmsian waltzes.
6. It was not until he was three years old that Einstein a) walked b) talked c) multiplied d) exhibited an interest in physics.
7. One of the fundamental postulates of the special theory of relativity is that a) the speed of light is constant b) the universe is finite c) the speed of light is infinite

d) there is an absolute zero of temperature.

8. Einstein was known to have been an accomplished musician. The instrument he played was the **a)** piano **b)** harp **c)** violin **d)** tuba.

9. Einstein's equations for the special theory of relativity are not used for most of the problems that are encountered under ordinary circumstances. They *must* be used, however, for problems that involve extremely **a)** high velocities **b)** low temperatures **c)** large distances **d)** rarefied matter.

10. In matters of dress, Einstein was more than a quarter of a century ahead of his time—a premature hippie, one might say. His getup did *not* include, however, **a)** a rope to hold up his pants **b)** long hair **c)** no socks **d)** a headband.

11. When Einstein published his special theory of relativity, he was **a)** a professor of physics at Berlin University **b)** a tutor of mathematics in a private home **c)** a scrivener in the dead-letter office **d)** a clerk in the Swiss patent office.

12. Bertrand Russell wrote a popular book on relativity called **a)** *Relatively Speaking* **b)** *The A B C of Relativity* **c)** *Fast Company* **d)** *Dr. Einstein's Magic Bullet.*

13. Einstein signed a famous letter to President Franklin D. Roosevelt. The letter **a)** warned of the dangers of Nazism **b)** led ultimately to the U.S. atomic bomb project **c)** asked him to establish the U.N. **d)** inquired about Roosevelt's health.

14.

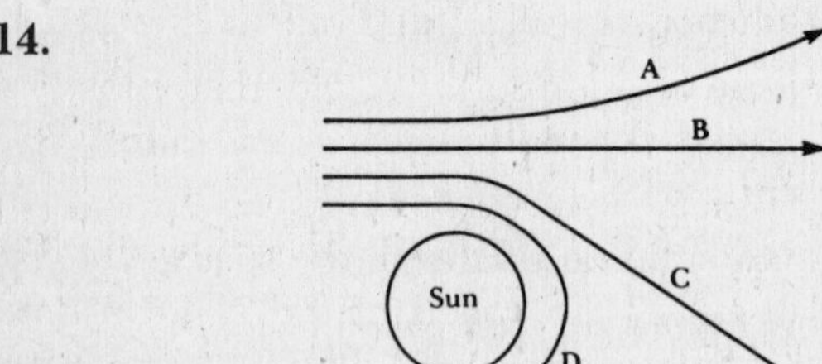

The general theory of relativity makes a specific prediction about the path of a light ray from a distant star, as it skirts the sun. The figure above is schematic; A and C are intended to represent paths with very slight curvature while B is a straight line. The prediction of the theory is illustrated by **a)** A **b)** B **c)** C **d)** D.

15. The success of the general theory of relativity with regard to the motion of the planet Mercury lies in the precise prediction of **a)** increased speed in certain segments of its circular orbit **b)** the speed as a function of location in its elliptical orbit **c)** the rate at which it spirals farther and farther out from the sun **d)** the rate of precession of its perihelion.

16. The National Academy of Sciences and the National Academy of Engineering commissioned a memorial to Einstein, consisting of a bronze figure of the famous scientist, twelve feet high, contemplating **a)** an exploding nucleus **b)** a bust of Homer **c)** a model of the universe **d)** a relative.

17.

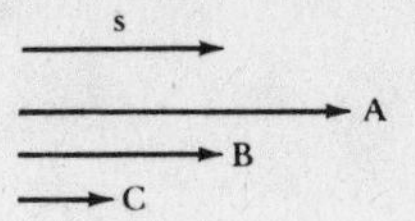

In the figure above, if *s* is the length of a rod as measured by an observer at rest with respect to the rod, the length as measured by an observer moving to the right at very high speed with respect to the rod will be **a)** A **b)** B **c)** C **d)** not measurable.

18.

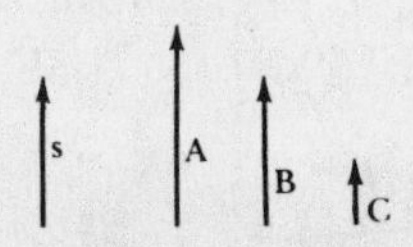

In the figure above, if *s* is the length of a rod as measured by an observer at rest with respect to the rod, the length as measured by an observer moving to the right at very high speed with respect to the rod will be **a)** A **b)** B **c)** C **d)** not measurable.

19. The twin paradox a) states that one cannot know two properties of matter simultaneously b) arises in the quantum theory of electromagnetic radiation c) arises in the general theory of relativity d) refers to the fact that Einstein had a twin brother with a low IQ.
20. An example of Einstein's work that is not as well known as his work on relativity is a) why rivers meander b) how to predict weather c) recycling waste d) the aging process.
21. One of the consequences of the theory of relativity is that

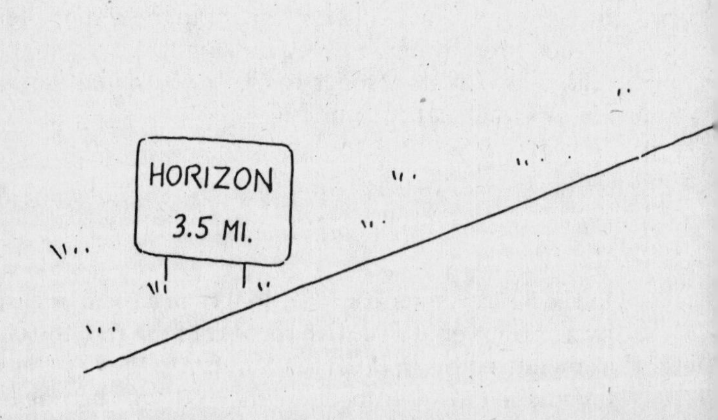

the duration of time intervals depends upon the speed with which one is traveling. A spaceship moving at high speed has a clock aboard. An observer in the spaceship —as opposed to a person on Earth—measures, on the clock in the spaceship, what is called **a)** solar time **b)** Greenwich mean time **c)** proper time **d)** three-quarter time.

22. *The Physicists,* a play by Friedrich Durrenmatt, includes in its cast of characters Einstein, Newton, and Möbius. The action takes place **a)** in the Middle Ages **b)** on the Isle of Elba **c)** in a dream **d)** in an asylum.

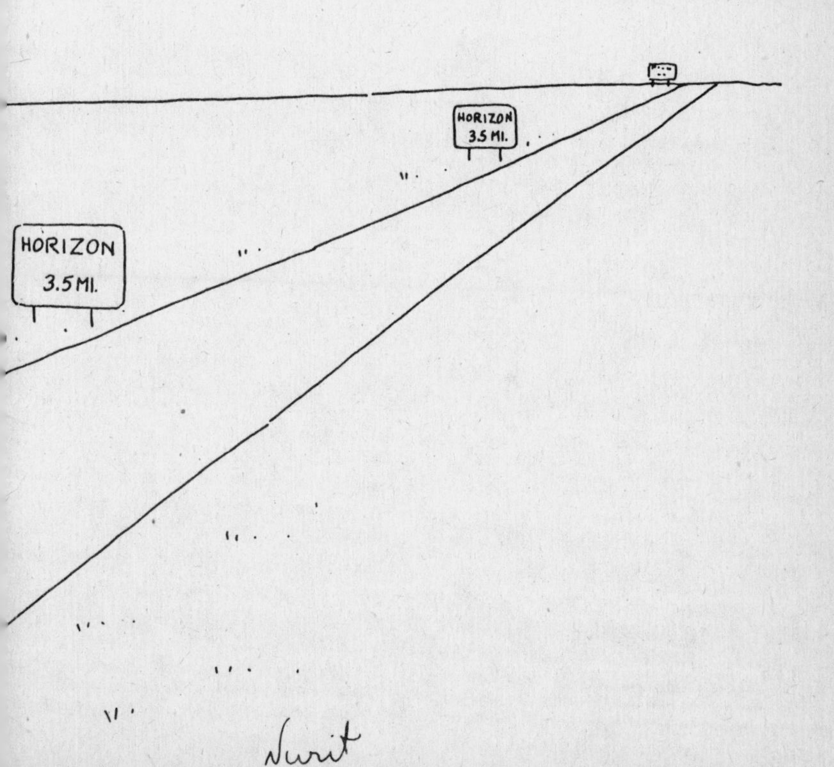

Answers

— • • • —

1

Do You Have a Fine 14th- or 20th- Century Mind?

1. b
2. b
3. a. A lunar eclipse is an eclipse *of* the moon.
4. c
5. c. Fusion is the process whereby light atomic nuclei combine. Thermonuclear energy is derived from fusion. For alpha decay, see the answer to question 11.
6. a
7. b. The word *month* comes from the length of time it takes the *moon* to orbit Earth.
8. a
9. b
10. c
11. a. Radioactivity consists primarily of alpha decay (the emission of an alpha particle, the nucleus of the helium atom), beta decay (the emission of a beta particle, an electron) and gamma emission (emission of an energetic electromagnetic "particle," or photon). The terms alpha (α), beta (β) and gamma (γ), the first three letters of the Greek alphabet, were introduced before the nature of the radiations was understood.
12. a. The Earth is believed to be about four and one-half billion years old; dinosaurs were around one hundred to two hundred million years ago, and man has been on Earth a few million years.
13. d. Helium's discovery on the sun gave it its name; the Greek word for sun is *helios.* (Note that the last statement in question 33 constitutes a heliocentric [sun-centered] theory.)
14. d
15. d
16. b
17. d. Little was known about electricity in Newton's time.

18. c
19. d. Deoxyribonucleic acid (DNA), the carrier of genetic information, is studied in molecular biology.
20. b. In a transmutation, one element becomes another.
21. c
22. b
23. a
24. d
25. b
26. b
27. d
28. c
29. b
30. d
31. d. The word *atom* comes from the Greek word meaning *uncuttable* or *indivisible.*
32. c. The probability that a particular radioactive nucleus will decay within a given time is independent of its history, and, in particular, of the time that has elapsed since it was created. The analogous statement for people is *not* true; the life expectancy of a person depends very much on the person's age, as is taken into account by insurance companies when they issue life insurance policies.
33. d
34. a. NaCl is table salt, IBM stands for intercontinental ballistic missile, MIRV stands for multiple independently targeted reentry vehicle, and U-235 is the isotope of uranium that has a total of 235 particles in its nucleus. SALT stands for strategic arms limitation talks.

• • 2 • •

Far Out

1. **c.** If the speed of light is taken to be approximately 200,000 miles per second (the actual value is 186,000), it took light about 1,000 seconds to travel from the satellite, near Mars, to Earth, and thus about 2,000 seconds—roughly 30 minutes—to travel from the satellite to Earth and back. That was far too long to permit Earth-controlled adjustments just before landing.
2. **b.** The sun consists primarily of hydrogen and helium; it contains few of the heavy elements required for fission. The radius is nearly constant; therefore, very little energy comes from changes in gravitational energy.
3. **d**
4. **a.** Mark Twain was born when the comet was visible and is said to have predicted—accurately—his death on its return 76 years after his birth. It's due again in 1986.
5. **c.** When you look at the sky, during the day, in a direction other than in the direction of the sun, the only light you see is scattered light. Because blue light is scattered more than red light the sky appears blue.
6. **c.** Mercury, Uranus, Neptune, Pluto. If we remove the restriction "directly," then, others can be added. For example, Thor, after whom thorium is named, is sometimes identified with Jupiter, and Vanadis (vanadium) is the Norse goddess Freya, sometimes associated with Venus.
7. **c.** This is a combination of letters set by running a finger down the first and then the second left hand vertical bank of keys of a linotype machine to produce a temporary marking slug not intended to appear in the final printing. However, this line is sometimes seen in newspapers, in sections that are obviously messed up. The combination bears some relation to the frequency of letters used in English, usually given as ETOANIRSHDLUCMPFYWGBVKJXZQ—though in his story "The Gold-Bug" Poe gives the sequence as EAOIDHNRSTUYCFGLMWBKPQXZ (only twenty-four letters, no J or V.)

UHURU, the Swahili word for freedom, is the name of a satellite launched from Kenya. The X's in *b* and *c* indicate that these are X-ray emitters.

8. **c.** Black holes can be detected indirectly by their gravitational effects.
9. **a.** The universe consists overwhelmingly of hydrogen and helium.
10. **a.** It is a very difficult experiment to perform, as the light rays bend by only 1.75 *seconds* of arc.
11. **b.** It is the temperature at the surface of the sun that determines the region of the spectrum that has the greatest intensity, and, therefore, the "color" of the sun.
12. **b.** It is the high temperature and the enormous number of possible proton pairs which make the fusion process possible in the center of the sun; the density at the center is high, but not extraordinarily high.
13. **d.** The farther the planet is from the sun, the longer the period. More precisely, the square of the period is proportional to the cube of the radius of the orbit. This is Kepler's third law, sometimes called by students the Times Square law.
14. **a.** The stars are Mizar and Alcor, in the handle. (Actually, Mizar consists of four stars.)
15. **d**
16. **c.** If you compare the two equations $v = Hd$ and $d = vT$, you see that $T = 1/H$. Thus, if you didn't already know the approximate age of the universe from newspaper reports, you could have deduced the correct choice.
17. **b.** One can now detect radio waves bounced off the moon and Venus, but not off stars.
18. **a**
19. The first is Neptune's trident; the male symbol also stands for Mars; the female symbol is also used for Venus; and the last is the symbol for Earth. These symbols are used by astrophysicists as subscripts to planetary data. For example, the mass of the Earth is denoted by $M_{\oplus}$.
20. Mars and Milky Way are among the better known. There are also Galaxy, Starburst, Stars, Starlight Mints. . . .

••3••

Made in U.S.A.

1. d
2. **b.** 454 *grams* weigh one pound. This follows from the correct answer and the fact that there are 1,000 grams in a kilogram. There was a relation between shillings and pounds, but not through weight; in the old British monetary system, there were twenty shillings in a pound.
3. **a.** It is British. The transistor was invented by John Bardeen, W. H. Brattain, and W. B. Shockley; the cotton gin (a shortened version of the word *engine*) by Eli Whitney; and the zipper by Whitcomb L. Judson.
4. **d.** Thomas Edison gets the credit. Holography was invented by Dennis Gabor, a Hungarian-born Britisher; there were many contributors to what is now called photography, the early ones all European; and the jet engine for airplanes is British.
5. d
6. b
7. **c.** And Franklin didn't even get his feet wet. Shipping schedules indicating that ships took less time to cross the Atlantic along certain routes confirmed for him what he had suspected from the appearance of the ocean while voyaging. He had a Nantucket whaler, who, incidentally, was his second cousin, take temperature and speed measurements to help map the stream. These measurements were mostly on the North Atlantic Drift, which is the continuation of the Gulf Stream and often confused with it. The actual discoverer of the Gulf Stream was Ponce de León, in 1513.
8. **a.** Electromagnetic induction is the production of an electric field by a changing magnetic field or by a conductor moving in a magnetic field.
9. **d.** Brookhaven is on Long Island, Oak Ridge in Tennessee, and Argonne near Chicago. Marne is the name of a terrible World War I battle, in France.
10. **d.** If an American invention is not attributable to Edison, a

good guess is Franklin. A remarkable fellow. Mozart, Beethoven, and Gluck composed pieces for the glass harmonica and Marie Antoinette played it. The eighteenth-century German physician Franz Anton Mesmer used it as background music for his experiments with hypnosis (observed by Franklin in Paris) because it tended to lull his subjects into a trance. Further fascinating details can be found in M. E. Grenander's "Reflections on the String Quartet(s) Attributed to Franklin" (*American Quarterly,* March 1975). Bora Minnevitch was an expert on the ordinary harmonica.

11. b

12. a. Lawrencium, Lr, is element 103 in the periodic table of the elements. Lawrence invented the cyclotron.

13. c. Berkelium, Bk, is element 97.

14. b. Californium, Cf, is element 98. At the time lawrencium, berkelium, and californium were found, the accelerator at Berkeley, California, was essentially the only one in the world in which these transuranic elements could be produced; therefore there appears to be an element of greed—if not egotism—in their naming.

15. a. Americium, Am, is element number 95 in the periodic table of the elements.

16. d

17. c. This is a fundamental postulate of the special theory of relativity. It is not clear whether Einstein knew of the experiment when he formulated the theory.

18. c. Watson shared the Nobel prize in physiology or medicine in 1962 with F. H. C. Crick and Maurice Wilkins for their discovery of the molecular structure of DNA.

19. c. Michelson and Rabi were born in what is now Poland; Lee was born in China.

20. a. Pauling won the Nobel prize in chemistry and in peace. Urey and Seaborg won chemistry prizes, Roosevelt the peace prize.

21. a. For the transistor and for the theory of superconductivity.

22. c. Marie Curie won the Nobel prize in physics and in chemistry. Roentgen received a physics prize for the discovery of X rays, Rutherford a chemistry prize for work in radioactivity, and Pavlov the prize in the physiology or medicine category, for work on the physiology of digestion. Incidentally, have you ever wondered why there is no Nobel prize in mathematics? The prizes are awarded in the fields of chemistry, physics, physiology or medicine, literature, and peace. (Economics is a recent addition and was not specified in Nobel's will.) The reason is not a very scientific one. The story goes that Alfred Nobel was interested in a woman who went off with somebody else, the mathematician M. G. Mittag-Leffler. Presumably, if there had been a mathematics prize, M. G. M.-L. would have been a good bet to receive it. Nobel's enmity toward one mathematician was generalized to the set of all mathematicians.

23. c

24. d

25. b. Scopes was the name of the teacher brought to trial. The Scottsboro trial involved a tragic miscarriage of justice in the South, Snopes is better known to William Faulkner fans than to jurists, and Jukes was the inbred family studied intensively by psychologists.

26. Jefferson—agriculture. Grant—mathematics. Hoover—engineering. Nixon—electronic devices (especially tape recorders).

••4••

Misfits

1. **c.** Galen was an ancient Greek physician. Tycho Brahe (1546–1601) was a distinguished Danish astronomer whose data were used by Kepler to determine that the planets move in elliptical orbits. Giordano Bruno was an Italian philosopher, born in 1548, whose career was terminated in 1600—he was burned at the stake—because of his cosmological beliefs. Galileo's dates are 1564 and 1642.
2. **d.** Thomson, an Englishman, was the discoverer of the electron, in 1897. Josiah Willard Gibbs (1839–1903), relatively unknown to the general public, is considered by many scientists to be the greatest American-born scientist, an assessment based primarily on his work in statistical mechanics. Joseph Henry (1797–1878), independently of Michael Faraday, discovered that a change in magnetic field could induce an electric field—the basis of generators; he was one of the founders of the National Academy of Sciences of the United States. Count Rumford was born Benjamin Thompson in the United States, and was made a count of the Holy Roman Empire by the Elector of Bavaria. He did important work on heat.
3. **a.** England. Choices *b, c,* and *d* are the United States.
4. **b.** Paderewski was "only" a concert pianist and a prime minister of Poland. De Valera, a prime minister of Ireland, taught mathematics, as did Lewis Carroll, an Oxford don.
5. **c.** For this principle, consult an economics text.
6. **c.** Robert Hooke, English physicist, mathematician, and inventor, lived roughly half a century before electricity became a "hot" field. He did contribute to the field of elasticity. (Hooke's law states that within the elastic limit, strain is proportional to stress.) The other three choices have electrical units named after them.
7. **d.** Proctology might be offered in a medical school; it is the study of the anus and rectum.
8. **d.** A radian is an angular unit. A mil is one-thousandth of an inch

and is used chiefly to specify the diameter of wire. A parsec is an astronomical unit. A light-year is the distance light travels in one year.

9. **b.** A silo is not a unit at all, although one could refer to it as a storage unit. A barn is 10^{-24} cm^2; it originated in nuclear physics, where an effective area of that size, within which a reaction occurs, is enormous—like the side of a barn.
10. **a.** Aeschylus was a playwright; the other three are mathematical greats.
11. **d.** A rational number can be written as the ratio of two integers. A perfect number is one whose factors, other than itself, add up to itself. A prime number is one that has itself and unity as its only factors. Choice is a grade of meat.
12. **d.** A specter is a ghost. A matrix is normally a two-dimensional array of numbers. A vector is a quantity completely specified by a magnitude and a direction. A lattice, in physics, is a regular, periodic configuration of points, particles, or objects throughout an area or space.
13. **d.** Surrealism is a literary and artistic movement that began in the 1920s. Sidereal has to do with stars or constellations. A sidereal day is the interval between two successive transits of a star over the meridian. A solar day is the interval between two successive meridian passages of the sun. Synodic relates to the period between two successive conjunctions of the same celestial bodies, such as the sun and moon.
14. **b.** A peristylium is a colonnade surrounding a building or court. In the present context, the other terms are all related to orbital motion, the perihelion being the point in the orbit nearest the sun, perturbation being disturbance of the orbit, and precession being, for Mercury, a slow rotation of the orbit in its plane.
15. **c.** Not at the time of this writing, anyway.
16. **c.** Antoine Capon was specially created for this occasion. John Cabot, who sailed under the English flag, was born Giovanni Caboto, in Genoa, Italy.
17. **b.** The phlogiston theory postulated that inflammable materials gave off phlogiston in burning. The geocentric theory, as the word implies, had the Earth at the center of the universe; and the ether theory postulated an all-pervading, infinitely elastic,

massless medium as necessary for the propagation of electromagnetic waves (such as light and radio waves).

18. **d.** (However, Wolfschmidt did give his name to an alcoholic beverage, which *some* might regard as a constant of nature.) Avogadro, an Italian, stated that all gases at a given temperature contain the same number of particles per unit volume. His name is given to the number of atoms or *molecules* (a word Avogadro coined) present in an amount of substance that has a mass in grams numerically equal to its atomic or molecular weight. Loschmidt's number is the number of molecules per unit volume of an ideal gas at 0°C and normal atmospheric pressure. Boltzmann, an Austrian physicist, noted primarily for his work in statistical mechanics, was one of the all-time greats. The molecular gas constant is named for him.
19. **d**
20. **a.** Toynbee was a historian. Rutherford performed the first nuclear transmutation, Pauli was the author of the exclusion principle, and Bohr was the father of what is now known as the old quantum theory.
21. **b.** However, Pauling has enough other things to his credit.
22. **b.** $225 = 15^2$ and $1024 = 32^2$.
23. **a.** *Compulsive* might be found in a psychology text. (Compulsive authors might feel compelled to define *compulsive.*)
24. **b.** Baron de Kalb was a Revolutionary War hero. DeKalb is also a stop near Fulton on subway lines in Brooklyn. (We trust we will be forgiven a bit of provincialism.)
25. **b.** *Per stirpes* is a legal term relating to dividing inheritances according to branches of the family.
26. **b.** The binary system, a number system that has *two* as its base, is used in computers; base *three* is not. Hardware denotes the mechanical and electrical devices from which a computer is constructed. Software refers to some computer programs.

• • 5 • •

Designer Genes

1. c
2. c
3. d. If we estimate humans to be *roughly* 100 cm × 50 cm × 10 cm = 50,000 cm^3 in volume, we can eliminate *a* and *b*, which would make cells on the average 1 cm and .1 cm on a side, respectively.
4. b
5. b. Choices *a, c,* and *d* represent back, blood, and breath problems, respectively.
6. d. Normally, females are XX and males XY. Whether an extra Y chromosome in men is linked to a high degree of aggressiveness has not been established.
7. b. Huskies frequently have one blue and one brown eye. Although elementary genetics texts often give eye color as an example of simple Mendelian inheritance, it is now known that the inheritance mechanism for eye color is more complex.
8. d
9. c. The gene combination possibilities are brown-brown, brown-blue, blue-brown, or blue-blue; only in the last case will the child have blue eyes. (See the comment in the answer to 7.)
10. c. Choices *a, b,* and *d* are associated with chicanery of a non-genetic kind. Potemkin was a field marshal and favorite of Catherine II, alleged to have built sham villages (Potemkin villages) for Catherine's Crimean tour; Rasputin was a "holy man" at the court of Nicholas II, with a hold over the rulers; Chichikov is the protagonist in Gogol's *Dead Souls.*
11. b. A child has half a chance of having one gene for blue and one for brown, and half a chance of having two genes for blue. (See the comment in the answer to 7.)
12. d. Modern medical treatment has modified the statement in *a* by prolonging lives.
13. a. Borlaug's prize was for peace.
14. a. $\frac{1}{5} \times \frac{1}{5} = \frac{1}{25}$. To have the recessive version of the character-

istic, the individual must have both genes recessive; the probability of having a given recessive gene is 1/5.

15. **b.** One meaning of homogamous relates to reproduction within an isolated group. Homeopathic pertains to a system of medical practice.
16. **a.** Somatic comes from *soma,* Greek for "body." In *d* we were capitalizing on the similarity in sound of words relating to sleep, which come from *somnus,* the Latin for "sleep."
17. **b.** But after the atomic bomb came into being, some radioactive fallout did end up in milk.
18. **a.** True, Down's syndrome was originally called mongolism, but not for the reason given in *d;* it was so called because the eyelids of afflicted persons have a slanting fold, suggestive of the eyes of Orientals.
19. **d.** They do include adenine.
20. **b.** Since there are many genes in each of the 23 pairs of chromosomes, *a* can be eliminated. Choices *c* and *d* might be guessed to be incorrect on the ground that, while human beings have many characteristics, there would be little "point" in having billions or even millions of genes.
21. **c**

••6••

Principia Newtonia

1. **a.** The unit of force in the "barbaric" system, properly called the British system of units (foot, pound, second), is the pound. In the cgs system (centimeter, gram, second), the unit of force is the dyne. In the mks system (meter, kilogram, second), the unit of force is the newton. One newton equals 10^5 dynes. The cookie, the Fig Newton, was apparently named not after the scientist but after a district in Massachusetts.
2. **b.** Newton invented the first reflecting telescope, that is, a telescope in which the principal focusing element is a mirror rather than a lens, as in the refracting telescope.
3. **c.** The acceleration due to gravity at the Earth's surface is almost always denoted by g, not G.
4. **d.** Since force = mass × acceleration, the pull of gravity is proportional to the mass. The accelerations of the cannonball and the pebble—and, therefore, their final velocities and the time of fall—will be the same (if we neglect the very small effect of air resistance).
5. **b.** This is a consequence of the conservation of angular momentum. The speed of rotation—angular velocity—must increase to compensate for the decrease in the moment of inertia.
6. **a.** Newton is credited by some with the establishment of the gold standard.
7. **d**
8. **b.** There are no conservation laws for *a* and *d.*
9. **c.** Given initial data, one can, in principle, predict everything. The famous mathematician Laplace stated, in 1814, that, for an intelligence which could comprehend all the forces of nature, if given at any instant the conditions of each of the constituents of the universe, "nothing would be uncertain and the future, as the past, would be present to its eyes."
10. **b.** Incidentally, molecules were not known in Newton's day. The covenant associated with the rainbow was between God and Noah; God agreed never again to destroy human beings by

a flood and set the rainbow in the sky as a sign of the covenant.

11. d. At this he was—as might be expected—one of the greats. As for womanizing—Newton is believed to have died a virgin.

12. a. Again, Newton's work was outstanding. Haruspicy is divination; in ancient Rome predictions were based on inspection of entrails of sacrificial animals.

13. d. Choice *b* requires quantum theory, a twentieth century development. Neither *a* nor *c* falls in the domain of mechanics, the domain covered by Newton's laws.

14. d. Woodrow Wilson, for example, in *Constitutional Government in the United States,* says that the government was constructed on the Whig theory of political dynamics, which was copied from Newton's theory of the universe.

15. a. The wavelength is about 6×10^{-5}cm.

16. a. The densities of the sun and Earth are about 1.5 and 5.5 gm/cm^3, respectively.

17. c. The density of the moon is 3.3 gm/cm^3; the density of Earth is 5.5 gm/cm^3. They are just close enough—or far enough apart—for controversy. Scientists who think the moon came from the Earth—possibly having been torn away by a passing star, or simply having split off—find the densities close. Scientists who hold that the moon either was assembled from satellites captured by the Earth, or was captured as a whole by the Earth, see the densities as far apart. The abundances of the elements and the isotope distribution in the moon rocks studied thus far are similar to those found in the Earth's mantle, findings at least consistent with the moon's having come from the Earth. That the moon is not made of green cheese is less controversial, as the density of green cheese—or, at least, that of the blue cheese we substituted for hard-to-find green—is rather close to one (determined by immersing it in water and noting that it sank slowly).

18. b. Newton and Leibniz were bitter rivals over priority in the invention of the calculus.

7

Letters from Greece

After the word for each Greek letter, the letter itself is given in upper- and lowercase.

1. rho Ρρ The visual purple is rhodopsin, the shrub is rhododendron, and a horizontal set of numbers in a rectangular array—called a matrix by mathematicians—is a row.
2. tau Ττ The symbols for tantalum and uranium are Ta and U, respectively. The Latin for bull is taurus.
3. epsilon Εϵ It sounds like upsilon.
4. pi Ππ A transcendental number is one which cannot be the root of an algebraic equation such as $3x^2 + 5x - 4 = 0$. The base e of natural logarithms is another example.
5. eta Ηη The prescribed forms make up etiquette. As a suffix: a packet is a small pack, for example.
6. beta Β β The light particle emitted in certain radioactive decays was named the beta particle; only later was it realized that beta particles are identical to electrons orbiting in atoms.
7. alpha Αα Just as with the beta, the alpha particle was named before its nature was understood. The river is Alph and the soup, of course, is alphabet.
8. nu Νν The important word is nucleus. The Russian and Yiddish for *Well?* transliterated, is *Nu?*
9. zeta Ζζ The Riemann zeta function involves sums of reciprocals of powers of integers.
10. psi Ψψ The word is psychology; the letter is epsilon.
11. lambda Λλ The Russian for yes is da.
12. phi Φϕ A vector in three dimensions can be defined by its length, the angle it makes with the z axis (the polar angle, often denoted by θ), and the angle that the projection of the vector on the xy plane makes with the x axis (the azimuthal angle, often denoted by ϕ). *Azimuth* is assumed to come from the Arabic for "direction." Jack's giant said, "Fe fi fo fum . . ."
13. delta Δ∂ The Dirac delta function vanishes everywhere except

at one point, where it is infinite. The muscle is the deltoid.

14. omicron Οο The precursor is Cro-Magnon.
15. gamma Γγ The factorial, defined by $n! = n(n-1)(n-2) \ldots 3(2)(1)$ is meaningful only for n an integer. The gamma function is an extension of the factorial to the case for which n is not an integer. George Gamow, an outstanding physicist, wrote some extremely entertaining books on physics and mathematics, including *Mr. Tompkins in Wonderland.* The game is backgammon.
16. omega Ωω The elementary particle is the omega minus, written Ω^-. Its existence and some of its properties were predicted by a theory of elementary particles, "The Eight-fold Way"; seven particles of a certain type were known, but the theory demanded an eighth. The discovery of the Ω^- gave an enormous boost to the theory. Gell-Mann, an American, is a Nobel laureate who teaches at the California Institute of Technology; Yuval Ne'eman is a noted Israeli physicist who has been associated with the University of Texas. It was once remarked about him (by Professor William Fowler) that in Texas, Yuval Ne'eman is known as You All Ne'eman.

    Mega, as a prefix, denotes one million, and *Little Women*'s Meg might have been O'Meg had she been Irish (and had we been allowed some license).
17. sigma Σσ The symbol Σ often denotes a sum. The society is Sigma Xi.
18. mu Μμ As an example, a microfarad, symbolized μF, is one millionth of a farad. Cats mew. (They also meow, but that doesn't sound like the Greek letter.)
19. iota Ιι The expression is "I ought to," said as "I oughta"; the car is the Toyota.
20. theta Θθ The place is a theater, and hypothetical is sort of close to imaginary.
21. chi Χχ The word is *chiasma* and the city Chicago.
22. upsilon Υυ The storage cylinder is a silo.
23. xi Ξξ The instrument is the xylophone.
24. kappa Κκ ΦΒΚ. Decapitated.

The letters in their proper order are: Α Β Γ Δ Ε Ζ Η Θ Ι Κ Λ Μ Ν Ξ Ο Π Ρ Σ Τ Υ Φ Χ Ψ Ω

••8••

Two Cultures

1. **d.** The term was taken by Murray Gell-Mann, the quark's co-postulator, from Joyce's line, "Three quarks for Muster Mark." A snark is a chimerical animal created by Lewis Carroll for *The Hunting of the Snark; sark* is a word we thought we made up, but it turns out to be a Scottish word for shirt; and *narc* is the street word for a narcotics agent.
2. **a.** Just as protons, electrons, and neutrons, being constituents of atoms, are more "basic" than atoms, so quarks are considered to be more "basic" than protons or neutrons, each of which is presumably composed of three quarks. Quarks have not actually been found; many physicists who believe they exist think they cannot be found because they cannot be isolated.
3. **c.** *Son et lumière,* often seen on French tourist ads for public monuments, means "sound and light" (shows).
4. **d**
5. **a.** *Inherit the Wind* was written by Jerome Lawrence and Robert E. Lee. *The Wind in the Willows* is a classic for children by Kenneth Grahame, *Trial by Jury* is a Gilbert and Sullivan operetta, and *Twelve Angry Men* is a film directed by Sidney Lumet.
6. **b.** The character is Levin, who resembles and apparently speaks for the author. *The Tomato Eaters* is a title concocted by us for this quiz, but Lysenko was real: he was a geneticist who claimed plants could inherit acquired characteristics—which, if true, would have been invaluable to agriculture. Lysenko had Stalin's support in his "research." Many of the leading geneticists in the U.S.S.R. were imprisoned, and Soviet genetics was set back for years. Although physicists under Stalin had to subscribe formally to a certain amount of nonsense—they were to reject the uncertainty principle, a fundamental tenet of quantum mechanics, for example—physicists as a group and physics itself did not suffer.
7. **b.** The piece is "The Moonlight Sonata."
8. **b.** "Claire de Lune" is Debussy's famous piece.
9. **d**

10. **c.** Tuscany is a region in Italy. Brahe was a noted Danish astronomer. Bruno, a philosopher, was burned at the stake in 1600 for stating that the universe is infinite. Torricelli invented the barometer.
11. **b.** The impressionists used small strokes of primary colors to simulate actual reflected light. Impressionism, expressionism, and fauvism were movements in art; chiaroscuro, from the Italian for light, *chiaro,* and dark, *oscuro,* is a representation in terms of lights and darks rather than colors.
12. **b**
13. **c.** In the novel, books are burned.
14. **b**
15. **b.** The computer in Arthur C. Clarke's and Stanley Kubrick's screenplay is known as Hal. *The Age of Uncertainty* is a book on the history of economics and economic theory, by John Kenneth Galbraith; *The End of the Road* is a novel by John Barth; and *The Money Game* is by the modern Adam Smith, an investment analyst who writes under that name.
16. **a.** The title of the work has suggested some humorous twists. A newspaper is alleged to have carried a photograph of Aristotle Onassis seeking a Hollywood home and considering one that had belonged to Buster Keaton. The photograph was captioned, "Aristotle contemplating the home of Buster."
17. **c.** Sulfur is the fuel of hellfire.
18. **a.** Vivaldi wrote *The Seasons,* also called *The Four Seasons.*
19. **c.** The density of platinum is 21.5 grams per cubic centimeter.

••9••

The Women's Questions

1. c

2. c. Androgyny, formed of *andro* (man) and *gyn* (woman) is the condition of having characteristics of both sexes. Misanthropy is a hatred of mankind. Ontogeny is the course of development of an individual organism.

3. d. The Galapagos Islands were Darwin's territory. Margaret Mead's books include *Coming of Age in Samoa, Growing Up in New Guinea, Male and Female,* and *Sex and Temperament.*

4. d. Fria was the Germanic goddess of love. Tuesday is named for Tiw, god of war, Wednesday for Woden or Odin, and Thursday for Thor.

5. b

6. a. A man has XY chromosomes. Therefore, a woman contributes one X chromosome to her child, while a man contributes either an X or a Y, so that, contrary to primitive—and sometimes not-so-primitive—belief, the woman is not responsible for the sex of the child.

7. a. Oveta Culp Hobby was Secretary of Health, Education, and Welfare under President Eisenhower; Margaret Thatcher, trained as a chemist, received plenty of recognition, but as England's prime minister; and Jocelyn Bell, as a student, called to Anthony Hewish's attention a curious effect—the first pulsar (ever discovered).

8. a. According to a geneticist, it was difficult to get statistics on baldness in women because those who were bald wore wigs—until someone thought of doing a study of women in a mental institution.

9. b. Color blindness is due to a recessive X-linked gene.

10. b. Professor Wu was a member of the group that provided the first experimental evidence of the nonconservation of parity.

11. a. *Patterns of Behavior* is a made-up name; *Life is with People,* by Mark Zborowski and Elizabeth Herzog, is a book about the culture of the shtetl; and *The Lives of a Cell* is the widely acclaimed book by Lewis Thomas.

12. **d.** Professor Burbidge is now at the University of California at San Diego. Tuesday Weld is an actress, Toni Morrison is an author (*Song of Solomon* is one of her books), and Anna Russell is a satirist of music.
13. **a**
14. **b.** Alternative *a* was inspired by Mozart's *Eine Kleine Nachtmusik, c* is due to Wordsworth in "My Heart Leaps Up," and the Klein bottle is a mathematical construct—a conceptual extension of the Möbius strip—in which there is no distinction between the inside and the outside.
15. **d**
16. **b.** The Nobel prizes are awarded for peace, literature, physiology or medicine, physics, chemistry, and (recently instituted) economics.
17. **a.** Euler was director of the Academy of Sciences in St. Petersburg. His first invitation to Russia came from Catherine I.
18. **a.** Karen Black is an actress, Karen Donner is invented, and Karen Blixen is the writer Isak Dinesen.
19. **b**
20. **a.** Eve was the writer daughter of Marie. Maria Goeppert Mayer, a Nobel laureate, was one of the developers of an important model of the nucleus. Lise Meitner and her nephew Otto Frisch were the first to understand that experiments by Otto Hahn and Fritz Strassmann represented nuclear fission. Roslyn Yalow, though a Nobel laureate in medicine, was trained in physics.
21. **d**
22. **b.** Rubella is also known as German measles. Salmonella and botulism are forms of food poisoning and roseola infantum is a disease of infants and children.

• • 10 • •

Going Public

1. **b.** Thalidomide was used in Europe; it led to the birth of thousands of badly malformed children.
2. **c**
3. **c.** Judge Kaufman was the judge in the Rosenberg trial; Professor Smyth was the author of a famous report on atomic energy.
4. **b.** Fermi led the group that produced the first controlled chain reaction, at the University of Chicago. Though they played a role in the atomic bomb project, Teller and Ulam are usually identified as the fathers of the H-bomb.
5. **a.** E. B. White is a writer, but not on science and government, and he's not British; he was associated with the *New Yorker* magazine in its early years. Snowden is the former husband of Princess Margaret. Desmond Morris was head of the London Zoo and wrote, among other books, *The Naked Ape.*
6. **c** (Office of Management and Budget). ABM refers to an anti-ballistic missile, IBM to an intercontinental ballistic missile or to the International Business Machines Corporation, and CBS to the Columbia Broadcasting System.
7. **b** (*roughly* 300,000 doctors in a population of two hundred million). A 40-hour-per-week work load gives 3 hours per year per patient.
8. **d.** Emma Goldman was an anarchist, who, incidentally, was jailed in 1916 for publically advocating birth control. Emma Lazarus was the poet whose sonnet is on the base of the Statue of Liberty. Flora Gibson is a name invented for this quiz.
9. **a** (Environmental Protection Agency). OPA was the Office of Price Administration. IND and BMT *should* be concerned with noise abatement—they are New York City subway lines.
10. **c.** "Nine old men" was the phrase applied by some to the justices of the Supreme Court at the time President Franklin D. Roosevelt tried to "pack the court."
11. **d.** The name "Plowshare" comes from the words of the prophet Isaiah: "And they shall beat their swords into plow-

shares, and their spears into pruning-hooks; nation shall not lift up sword against nation, neither shall they learn war any more."

12. a. The conferences are named after the site in Nova Scotia where the first one was held.

13. c. Actually in 1901.

14. c. *Roughly* 50 billion dollars out of 2,000 billion, the bulk going to development.

15. a. *Conseil Européan pour la Recherche Nucléaire.*

16. a

17. b. All were rulers in France. Nicknames were evidently a more convenient method of identification than numbers, for some purposes. The nickname for Charles the Simple apparently referred to his inability to deceive.

18. a. The act was passed during Lincoln's administration.

19. d. *New Scientist* is a British journal, semipopular in its coverage. *The Physical Review* is the primary outlet in the United States for research publications in physics. *Daedalus* is the journal of the American Academy of Arts and Sciences.

20. b. The minute hand is moved ahead or back as a function of international events relating to disarmament.

21. c. Ivan the Terrible was the first Russian ruler to bear the title *tsar.* Ivan Skivitsky Skivar is the hero of a song popular with scout troops (among others). Ivan the Critical is someone who doesn't exist but probably should.

••11••

Picture Show

1. **L, c.** The instrument of war was the atom bomb. Einstein's general theory of relativity predicted the bending of a light ray by a gravitational field; the bending could be seen only during an eclipse.
2. **B, a.** President Franklin D. Roosevelt was crippled by polio. Salk developed a polio vaccine; similar work was done by Albert Sabin.
3. **C, d.** Mendel studied properties of successive generations of peas.
4. **K, e**
5. **F, f**
6. **G, h**
7. **M, g.** The Carlsberg Foundation partly supported the Bohr Institute.
8. **I, m**
9. **E, l.** Galileo suffered house arrest for giving proof of Copernicus's heliocentric theory, and Giordano Bruno *was* burned at the stake for "extending" the theory.
10. **H, j.** Sakharov has been exiled to the remote town of Gorki, where he valiantly tries to keep up with, and do, research in physics.
11. **J, i.** *Cogito, ergo sum* ("I think, therefore I am") was Descartes's statement. The Cartesian system of coordinates is named for him.
12. **D, b.** The *peanut butter* and jelly sandwich. Carver found hundreds of uses for the peanut.
13. **A, k.** (The picture of Archimedes is not, of course, a true portrait.) Among other things, he is said to have destroyed an enemy fleet by focusing the sun's rays on the sails and setting the ships afire. He arrived at the principle of buoyancy while determining whether a crown was made entirely of gold; he weighed it in and out of water.

12

Head Hunting

1. **J, g**
2. **A, c.** In a right triangle, $a^2 + b^2 = c^2$, where c is the hypotenuse.
3. **G, i.** A comment on parity is included in the answer to question 10 on this quiz.
4. **C, j.** Paul Muni starred in the film *Dr. Ehrlich's Magic Bullet.* Salvarsan, or 606, was the six-hundred-sixth in the experimental series of tests for a treatment for syphilis.
5. **L, d.** Galileo discovered sunspots.
6. **K, f.** Darwin wrote *Voyage of the Beagle.*
7. **E, h.** The unit is the curie; the elements are, respectively, curium, polonium, francium, lutetium (Paris). Marie Curie and her husband Pierre were awarded the 1903 Nobel prize in physics for their study of the radiation phenomena that had been discovered by Antoine Becquerel, with whom they shared the prize. After Pierre's death, Marie won a second Nobel prize, in 1911, in chemistry, for her discovery of radium and polonium. Her daughter, Irène Joliot-Curie, and son-in-law, Frédéric Joliot-Curie, shared the chemistry prize in 1935 for synthesizing new radioactive elements.
8. **B, k.** Fermi's wife Laura wrote *Atoms in the Family,* which was published in 1954.
9. **I, b**
10. **F, e.** The prize was shared with C. N. Yang. Nature's distinguishing between left- and right-handedness manifests itself in the following way. A free electron is characterized by both momentum and spin. (Spin is intrinsic angular momentum, somewhat akin to the angular momentum of the Earth due to its rotation about its own axis.) An electron emitted in a radioactive beta-decay process is normally very energetic; for a very energetic electron, the direction of rotation associated with its spin is the direction in which the fingers of the *left* hand curl when the left thumb points in the direction of motion of the

electron. This result came as a shock to physicists, for it had been assumed that half of all such electrons would have spins described by the curl of the fingers of the left hand and half by the curl of the fingers of the right hand—in other words, that nature did not "prefer" one direction over another. (One of the first to recognize the possible deep significance of whether or not nature so discriminated was Newton's "anti-buddy," Leibniz.)

11. **H, 1.** Francis H. C. Crick and James D. Watson worked together and shared the 1962 Nobel prize for physiology or medicine with Maurice H. F. Wilkins for their work on determining the molecular structure of DNA. Watson's book, *The Double Helix,* was published in 1968.

12. **D, a.** The first atom bomb was constructed at Los Alamos, New Mexico, during World War II. The Manhattan Project—a code name—was directed by Oppenheimer. Oppenheimer quoted the *Bhagavad-Gita* when he saw the fireball in the test explosion of the atomic bomb. In 1954, during what is known as the "McCarthy era," Oppenheimer was deemed "not a good security risk" and denied access to classified information; the reasons given included Oppenheimer's past leftist associations and his opposition to the construction of an H-bomb. Many hold that this was an outrageous act on the part of the U.S. government. Despite his breadth and brilliance, Oppenheimer's creativity probably did not match his critical faculties. In physics *per se* he may well be remembered largely for his work on neutron stars.

13

The Art of Science

1. **b.** They were written as an epitaph intended for Sir Isaac Newton.
2. **b.** Brecht, a social critic, wrote three versions of the play. It is interesting to trace the changes in his beliefs in these versions. In the first, Galileo is a hero. In the second, he's a *bon vivant* —Brecht had gone to Hollywood and had met Charles Laughton, for whom he tailored the role. By the time the third version was written, the atomic bomb had been dropped and Brecht viewed scientists (and Galileo) as villains.
3. **c**
4. **b.** Alchemy was an honorable pursuit, attracting, among others, Isaac Newton, who devoted considerable effort to it. (One wonders which current "scientific" pursuits will be viewed in the future in the way we view alchemy.)
5. **d.** The doctor in the group. Harvey, court physician to James I and Charles I, discovered the circulation of the blood in the human body, in about 1616.

   Herschel might have made the *best* model for Whitman's "When I Heard . . ." for, not only was he the most renowned astronomer of his time, perhaps best known for having identified the planet Uranus, but he was also a well known organist and music teacher.

   Halley's name brings Halley's comet to mind, and Hubble was the first to show—in the twentieth century—that the universe is expanding.
6. **c**
7. **a.** Roger Bacon (1220–1292) can also be thought of as promoting the scientific method in that he called for experimentation and mathematics centuries before they became essential to scientific advance—which leads us to promulgate a theory based on these two pieces of data, namely: All Bacons promote the scientific method.

   Francis Drake was an English admiral. Roger Baldwin was

one of the founders of the American Civil Liberties Union.

8. **d.** The novel is *Gravity's Rainbow* and the song, "Over the Rainbow."
9. **d.** Born in France in 1885, he was a member of the group of composers known as *Les Six.* He later became an American citizen.
10. **a.** Cyrano de Bergerac wrote *Histoire Comique des Etats et Empires de la Lune* in 1657.
11. **a.** The work was *20,000 Leagues Under the Sea.*
12. **d.** Holst's suite is called *The Planets.*
13. **d.** The protagonist is Nerzhin in *The First Circle.*
14. **c**
15. **a.** Leonardo devoted a great deal of time to scientific questions. There are remarks in his writings that indicate glimmerings of Newton's second law of motion. He is better known, perhaps, for what might be classed as engineering: imaginative machines, particularly flying machines, which he designed but did not construct. His anatomical drawings combined both his scientific and his artistic interests.
16. **b.** However, Verne should have consulted scientists. In *From the Earth to the Moon,* written in 1865, some two hundred years after Newton did his work, Verne exhibits a serious lack of understanding of Newton's laws of motion.
17. **a.** The piece is *Mikrokosmos.* According to the Big Bang theory, the universe was exceedingly small, in a sense, a microsecond after it was created.
18. **c.** The novel had the same name. Bela Lugosi was the Hollywood actor who starred in *Dracula.*
19. **d**
20. **c.** *The Double Helix* was James D. Watson's behind-the-scenes account of his and Francis Crick's Nobel prize winning work. *The Structure of Scientific Revolutions* was written by Thomas Kuhn. *Revolutions in Science* doesn't correspond to anything we know. *The Spiral Staircase* was a film.
21. **b.** As far as we know, the other titles are in search of an author.
22. **a.** Philip Glass and Robert Wilson wrote the opera with Einstein's name in the title. The other scientists have not been so honored.

23. Michael Crichton
A. J. Cronin
Arthur Conan Doyle
James Herriot (a veterinarian)
W. Somerset Maugham (studied medicine)
Jonathan Miller
Walker Percy
François Rabelais
William Carlos Williams

and, as pointed out to us by Peter Barna, John Foreman, Marget Pach, and Thomas Sebeok:

Sigmund Freud
Richard Gordon
John Locke
Maimonides
Felix Marti-Ibañez
Axel Munthe
Alan Nourse

••14••

Frags

1. **frEE Energy.** A term in thermodynamics that pertains to internal energy and has different specific meanings for physicists, chemists, engineers, and others.
2. **liGHT Pipe.** It conducts light along curved paths. The index of refraction of the material of which the pipe is made must be high enough for the light within the pipe to be reflected rather than refracted. An important application of the light pipe is in medicine; it permits a physician to examine, through body openings, certain organs within the body.
3. **logariTHMic.** In the most common logarithmic scales, numbers are expressed as exponents of 10 or *e.* A *klotz* (pronounced klutz) is a log or lump (of wood) in German. Klutz has come into colloquial English, via Yiddish, as a clumsy person, logs being not particularly graceful.
4. **faLL Line.** At any point on the mountain, the fall line is the line of steepest descent among all lines passing through the point.
5. **adiabatiC PRocess.** Contrast it with an isothermal process, one in which the temperature is held fixed.
6. **SCHRödinger.** Erwin Schrödinger published his equation in 1926. Schroeder is the *Peanuts* character.
7. **reST Mass**
8. **Q Value.** A frequency can never be perfectly sharp. The Q value gives the ratio of the peak frequency to the spread in frequency and is, therefore, a measure of the sharpness of the frequency.
9. **baLL Lightning.** Most reports of the phenomenon state that the ball is about one or two feet in diameter. Responsible scientists claim to have seen ball lightning move into and out of a room through an open window.
10. **aBSTRact**
11. **soFT Landing**
12. **THReshold**
13. **shoCK Wave**

14. **absoRPTion**
15. **biG Bang.** According to the Big Bang theory—supported by the majority of cosmologists—the universe was created in a tremendous explosion roughly fifteen billion years ago. You may have noticed that we used slightly different numbers, in other quizzes, for the age of the universe. That will teach you not to take these figures too seriously.
16. **baND Width**
17. **deeP SPace**
18. **piTCH**
19. **louDSPeaker.** A transducer is a device which converts energy from one form to another.
20. **recALL ALL ALLoys.** The preferred spelling of the instrument is ukulele. (Hawaiian: *uku,* flea; *lele,* jumping.) However, as ukelele is also acceptable *(Webster's Third New International Dictionary),* we exercised some lexical license in order to get the triplet of letters EL EL EL in ukELELE Lady. Webster says the flea and jumping probably refer to the nickname of the lively little British army officer, Edward Purvis, who popularized the instrument, which is of Portugese origin, in the 1880s. Another derivation came to us from Marilyn and Ira Lichton of the University of Hawaii; according to them the flealike leaping refers to the player's fingers.

15

The Old Math

1. d. A hundred sets of three zeros are added to the three zeros in 1,000.
2. a
3. b. It was coined by Milton Sirotta, nine-year-old (at time of coining) nephew of mathematician Edward Kasner.
4. d
5. b. A rational number can be expressed as the quotient of two integers, while an irrational number cannot. Supposedly, members of Pythagoras's school were so elated on discovering that there were irrational numbers that they sacrificed hundreds of oxen.
6. a. Choices *b* and *c* represent $\cos\theta$ and $\tan\theta$, respectively.
7. c. Note that $3^2 + 4^2 = 5^2$. An isosceles triangle has two equal sides.
8. a. The Pentagon in Washington is a five-sided building. A pentathlon is a contest in which athletes compete in five different events. A pentahedron is a solid with five plane faces. The Pentateuch comprises the first five books of the Old Testament.
9. d. (It is simply Algol spelled backward.)
10. c. 50×101. Don't get depressed thinking the trick would not have occurred to *you* as a child; many rank Gauss as the greatest mathematician who ever lived.
11. b
12. c
13. a. Circle in the Square is an off-Broadway theater in New York City.
14. a. The integral of x^n is $x^{n+1}/(n+1)$. Incidentally, there is a Laplace transform in mathematics.
15. c. Geo means Earth; the rest of the word is related to measure.
16. c

16

"A Jug of Wine, A Loaf of Bread— And Thou Beside Me Singing . . ."

1. **c.** Etiology deals with causes; it can denote the philosophy of causation or, in medicine, the science of the causes of diseases. Rheology is the study of the deformation and flow of matter. Without comment, we note that ethology is the study of the formation of human character, or the study of animal behavior.
2. **c.** Taking somewhat unrealistic shapes for roasts, one, 10 by 10 by 1 inch, would require less time than one 5 by 5 by 4 inches, although they would weigh the same. In the first case there is no point in the roast that is more than half an inch from the surface, while in the second the center is two inches from the surface. For heat to penetrate takes time.
3. **d.** "Night and Day."
4. **a.** Robert Koch established the bacterial cause of many infectious diseases and developed tuberculin as a test for tuberculosis. The other two choices are our corruptions of Bacchus and Dionysus—the former often identified with the latter, who invented wine-making—both being "corruptors" themselves, in that their worship was drunken and orgiastic.
5. **c**
6. **b.** Kate Smith's big hit was "When the Moon Comes Over the Mountain." In occultation, the moon is between the Earth and the astronomical object. Since the location of the moon as a function of time is known exceedingly well, the direction—but not the distance—of the astronomical object can be pinpointed.
7. **d.** 200 proof would indicate pure alcohol.
8. **d.** Earth, air, fire, and water are what the ancient Greeks believed constitute the universe.
9. **a**
10. **a.** Methyl alcohol is poisonous.
11. **b.** C_2H_5OH is the chemical formula for alcohol. H_2O_2 is hydro-

gen peroxide, H_2SO_4 is sulfuric acid, and CH_3COOH is vinegar.

12. c. Red and yellow have longer wavelengths, and brown is a mixture of colors.

13. a. About the lowest alcohol content of beer is 3.2 percent. Therefore, 12 ounces of beer contain at least .38 ounce of alcohol. One ounce of 100 proof whiskey contains 50 percent, or half an ounce, of alcohol. Since most beer has more than 3.2 percent alcohol (4 percent to 6 percent), the can of beer and the shot of whiskey are comparable in alcohol content.

14. a. Read your Old Testament.

15. c

16. c

17. b. "Springtime in the Rockies."

18. b

19. c. One liter per day for 50 years is about 20,000 liters (or about 20,000 quarts), which is about 5,000 gallons.

20. b.

21. c. "Wait 'Til the Sun Shines, Nellie."

22. b

••17••

Sound and Music

1. c. Sound requires a medium (such as air) through which to propagate. (There is no true void, even in outer space, but there are regions where the density is exceedingly low.)

2. c

3. b. An angstrom is 10^{-8} cm, an ohm is a unit of electrical resistance, and a parsec is an astronomical unit of length equal to about three light-years.

4. b

5. a. Located in Montclair, New Jersey.

6. b

7. a. Assume, for example, that a long string, initially at rest, is caused to vibrate at one end. If the end vibrates 5 times per second, and the wavelength of the wave sent out is 8 feet, the first wave produced will have moved 5×8 or 40 feet in one second; that is, velocity = 40 ft/sec = frequency (5 vib/sec) × wavelength (8 ft).

8. a

9. b. One of us (GMS), at age seventeen (the maximum frequency one hears falls off with age), was in a physics class in which the instructor kept increasing the frequency of an oscillator and at 19,000 cycles per second said, "No one in the class can hear this now." GMS and another student—out of a large class—said they *could* hear the oscillator—whereupon they were promptly nicknamed Fido and Rover.

10. b. The velocity of a wave along a string is determined by the tension in the string and by the mass per unit length of the string and is not affected by where the string is clamped. The wavelength, however, is halved. Since velocity = frequency × wavelength (see question 7), the frequency is doubled.

11. b

12. b. The longest pipes support the longest wavelengths for the vibration of the air. The velocity of sound in air is determined by the properties of air and will be constant for a given temperature and pressure. Since velocity = frequency × wavelength (question 7 again), the frequency decreases as the wavelength increases.

13. **a.** An increase (decrease) in the tension of the string increases (decreases) the velocity with which a vibration is propagated along the string and, therefore, for a fixed wavelength, increases (decreases) the frequency. (Question 7 once more.)

14. **b.** In the fundamental vibration of a string fixed at both ends, every point other than the end points undergoes vibration.

15. **c.** The wavelength is more or less fixed by the dimensions of the resonant cavity. As the velocity of the sound is increased by a decrease in the density of the gas in the cavity, the frequency is increased (7 again). Originally, we sloppily used the word *swallow* instead of *inhale* for what the tenor did with the helium. We were informed by H. H. R. Friederici that if the tenor were to swallow helium before starting to sing, the helium would wind up in the stomach and not in the lungs. The quality of his voice would, therefore, not be altered, except that the larger "air" bubble in the stomach would restrict his breathing and, in addition, his song might be interrupted by belching.

16. **b** For what voiceprints are, see question 17. Voiceprints *have* been accepted in some courts in the last few years.

17. **c**

18. **a**

19. **c.** An important consideration in designing a concert hall—in choosing materials for the walls, ceiling, floor, and seats—is to have reflection and absorption properties that will give a reverberation time of the order of several seconds. Too short and the music sounds dead; too long and the music is muddy from echoes that are still bouncing around while new notes are being played. From 1937 to 1954, Arturo Toscanini conducted the NBC Symphony, which had been formed for him. A special broadcasting studio was constructed—Studio 8H. The first concert in this studio featured Beethoven's *Eroica* Symphony, which begins with distinctive chords. The chords sounded awful—like gunshots! The designers of the studio had devoted much effort to it—but without the formal first night audience with its full dress suits and starched shirt fronts, which changed the acoustical properties of the room.

20. c

21. c. Since paroxysms are spasms or fits, the dedication is not unreasonable.

22. a. Absolute pitch is relatively rare, even among musicians.

18

Famous Formulas

1. **F, b.** The Cavendish Laboratory is part of Cambridge University, England. Sir Henry Cavendish measured G, the gravitational constant. F is the gravitational force between two objects of mass m_1 and m_2 a distance r apart.
2. **E, c.** As important as the equation is, it does not have an "official" name. The one used here is a descriptive name. Quantities p and q are the distances of the object and image, respectively, from the lens, and f is the focal length of the lens.
3. **K, h.** Galileo Galilei. The term s represents the distance covered in the time t when subject to a constant acceleration a, starting from rest. To use it to estimate the time to hit the water —neglecting air resistance—set a equal to 32 feet per second per second and s equal to the distance to the water from the point at which you jump. If $s = 200$ feet, t will be about 3.5 seconds.
4. **G, d.** The symbol $\hbar$ is Planck's famous constant h divided by 2π, x is a spatial coordinate, t is the time, V is the potential, and m is the mass of the particle under consideration. The "wave function," Ψ, a function of x and t, has no direct classical interpretation, and, in fact, its meaning was misinterpreted by Schrödinger.
5. **B, g.** Hersey's book was *Hiroshima.* U represents an isotope of uranium. If, on the average, more than one neutron, n, emerges from a capture of a neutron by uranium, there can be a chain reaction.
6. **A, e.** The coulomb is a unit of electrical charge; k is a numerical constant, the value of which depends upon the units employed. F is the force between two charges q_1 and q_2 a distance r apart.
7. **I, a.** The symbol m represents the slope of the straight line, and the intercept b is the height at which the line crosses the y axis, that is, the value of y at $x = 0$.
8. **L, k.** Symbols a and b represent the lengths of the semi-major and semi-minor axes, respectively.

9. **J, l.** The p-p (proton-proton) chain is a series of reactions in which protons are fused into alpha particles and enormous amounts of energy are released. D and T, the deuteron and triton, are isotopes of the proton; He^4 is an alpha particle (the nucleus of the helium isotope of mass 4), and n is a neutron.
10. **H, f.** It is concerned with the regular shifting of consonants in groups. The *p* and *t* in the Latin *pater* become the *f* and *th* in the English *father*, for example. Other examples are the initial sound of the Latin *decem*, which becomes the English *ten*, and the initial sound of the Sanskrit *dhar*, which becomes the English *draw*.
11. **D, n.** Quantity m is the mass, h is the height, and g is the acceleration due to gravity.
12. **C, i.** It contains a parameter λ, which determines the half-life, a concept which assumes history is irrelevant. Human beings do not have a half-life because the length of time they can be expected to live depends upon the length of time they have already lived. (Contrast the life expectancy of an infant with that of a 60-year-old.) N_0 is the initial number of radioactive nuclei, $N(t)$ is the number at time t. Half the nuclei decay in a time $\lambda \ln 2$, where ln2 is the natural logarithm of 2. The product nucleus is called the daughter.

••19••

Chemical Solutions

1. c. For an element to be found on Earth, it must be stable, or have a half-life comparable to or greater than the age of the Earth, or it must be produced continuously in some fashion. Elements with a nuclear charge greater than 92 do not have any of these properties. They include neptunium (93), plutonium (94), and einsteinium (99). As of 1981, the element with the highest nuclear charge that has been definitely identified, is element 106; its name is still a matter of controversy. (One name has been proposed by American scientists, another by Soviet scientists, each group claiming priority.)
2. a. Sulfuric acid. Isomers, in chemistry, are compounds that have the same number of atoms of the same elements and hence the same molecular formula, but differ in the structural arrangement of the atoms and hence in properties. Polymers are long-chain molecules consisting of repeating units.
3. d
4. c
5. a
6. a
7. c. Nobel prizes were instituted half a century after his death. Avogadro's number, 6.02×10^{23}, is the number of molecules in a mole (gram molecular weight). Rutherford, born in New Zealand, was perhaps the preeminent experimental nuclear physicist. He won the Nobel prize in chemistry for work on radioactivity; research on elements was then considered chemistry. In his acceptance speech, he said he had observed many transformations while working with radioactive materials, but none as rapid as his own, from physicist to chemist. Mme. Curie won both a chemistry and a physics Nobel prize. Arrhenius, born in Sweden, won his prize for his theory of ionic dissociation.
8. d. Halogens include fluorine, chlorine, bromine, iodine, and

astatine. Anions are negatively charged ions, the ones that migrate to the anode in an electrolyzed solution; cations are positively charged ions and migrate to the cathode. A halcyon is a bird from ancient legend, identified with the kingfisher, that nested at sea and calmed the waves. But the word is more likely to be found as an adjective meaning calm and peaceful.

9. a
10. b. A molecule of oxygen (O_2) has a lower energy than two atoms of oxygen.
11. b. The others were made up.
12. c. An ion is an atom or group of atoms that lacks or has gained one or more electrons.
13. b. That is the point of putting salt on snow in winter.
14. d. One calorie raises 1 gram of water 1°C. Vaporization requires many more calories; the exact value is 540.
15. d. Compounds of noble gases were only recently produced. (Alfred spelled his name Nobel.)
16. d
17. b. The pH is the negative logarithm of the effective hydrogen ion concentration on a scale of 0 to 14 with 7 representing pure water, or neutrality; values less than 7 represent acidity, values greater than 7, alkalinity. In other words, instead of saying that the concentration of hydrogen ion in pure water is 1×10^{-7}, one says the pH of pure water is 7.
18. c. Many carbon compounds were derived from living organisms.
19. b
20. d
21. b. Some points of correspondence are: 0°C = 32°F and 100°C = 212°F. On the Celsius scale, −273°, 0° and 100° are, respectively, the lowest possible temperature and the freezing and boiling points of water. To obtain the answer, recall that $C = (5/9)(F - 32)$ and set C equal to F.
22. a. Chlorine bleach, sodium hypochlorite, is NaOCl; natural gas usually contains methane (CH_4) and other gases; hydrogen sulfide (H_2S) has the odor of rotten eggs.

23. a. A free radical is a group of atoms that act as a single entity in chemical reactions.

24. b. Choice *c* is an adiabatic process.

25. c. (Seriously!)

••20••

Famous Figures

1. c
2. a
3. d. DNA. A Möbius strip, named for August F. Möbius, a nineteenth century German mathematician, is a one-sided figure. To form one, take a long strip of paper, hold one end fixed, rotate the other end through 180 degrees, then glue or staple the two ends together. The resulting Möbius strip, a circle with a twist in it, has a number of interesting properties. That it has only one side can be demonstrated by starting at one point and moving your finger along the strip; your finger will move along the "inside" and the "outside" before it returns to the starting point. Further, if you cut the strip along its length, you do not get two separate strips but two interlocked circles.
4. b
5. b. At constant temperature, for an ideal gas, $PV = K$, a constant; this is *Boyle's* law. The curve, on a graph of P versus V, is a rectangular hyperbola.
6. d. The expression $4\pi R^2$ is the area of the surface of a sphere; the area of a circle of radius R is πR^2.
7. a. The circuit shown is commonly called an LC circuit or oscillating LC circuit, L for inductance, C for capacitance. If the capacitor is charged and then connected to the inductance, as in the diagram, the capacitor will start to discharge and energy will flow from the capacitor to the inductance, then back and forth indefinitely, in the ideal case. Since in an actual LC circuit there is always some resistance, the oscillations will die away unless arrangements are made to supply enough energy from an outside source periodically to compensate for that lost.

   Feynman diagrams appear primarily in elementary particle theory.
8. c. Michael Faraday, a very down-to-earth physicist, introduced the scheme of drawing pictures to represent electric fields in order to provide a "feel" for the character of the field. The

number of lines emerging from a positive charge, or converging on a negative charge, is proportional to the magnitude of the charge. In the case illustrated here of equal but opposite charges, the number of lines emerging from the positive charge is equal to the number converging on the negative charge; no lines run out to infinity. The direction of the field at any point is defined by the direction of the lines in the immediate neighborhood; the magnitude of the field at a point is proportional to the density of lines in the immediate neighborhood. Very close to either charge, the effect of the other charge is negligible and the lines are radial.

Tokamak is a Russian acronym for toroidal current machine, a device which will be helpful, one hopes, in ultimately making thermonuclear power through fusion a reality.

9. c. A plane wave, represented by the straight lines on the left, incident on a slit, produces the circular pattern on the right. This can be seen with water waves.
10. a. $\vec{\mathbf{A}} + \vec{\mathbf{B}} = \vec{\mathbf{C}}$.
11. b. The equation $x^2/a^2 - y^2/b^2 = 1$ is the equation of a hyperbola. It has two branches.
12. d. There *is* a Poisson distribution—named after Siméon Denis Poisson, a French mathematician (1781 to 1840)—which is widely used. The figure shown is known under a variety of names including a Gaussian curve, a normal probability curve, and an error curve.

• • 21 • •

Homage to Albert

1. c
2. c. Nevertheless, he went on to get a higher education.
3. b. If observer A is at rest with respect to a rod which he measures to have a length *L*, and if observer B is moving with a velocity v relative to A along the line of the rod, B will find the rod to have a length equal to $L\sqrt{1 - (v^2/c^2)}$.
4. b. *Ein* is one in German; a stein is a mug. Dinny was Alley's dinosaur, Professor Moriarity was Sherlock Holmes's arch foe, and the *d* choice was inspired by the cartoon character Speedy Gonzales.
5. a. The botanist was Robert Brown. Brownian motion is sometimes referred to as a "random walk."
6. b. It has been suggested that this is the origin of Einstein's thinking in visual terms.
7. a
8. c
9. a
10. d
11. d. Choice *c* was the occupation of Melville's Bartleby.
12. b. *The A B C of Relativity* was issued in 1925. The others were issued for this quiz.
13. b. Drafters of the 1939 letter included Leo Szilard and Eugene Wigner. The irony of the fact that Einstein, a pacifist most of his life, is "responsible" for the atomic bomb has often been noted.
14. c. The figure is *very* schematic. The bending is only 1.75 seconds of arc for light incident at the sun's limb.
15. d. According to Newtonian theory, the orbit of a planet is an ellipse, which is retraced over and over. Thus, the perihelion, the point in the orbit closest to the sun, remains fixed. According to Einstein's theory, the orbit changes very slowly; one can think of the ellipse—and with it the perihelion—as slowly precessing. *Peri,* incidentally, is Greek for "near," and *helios* is the Greek word for the sun.

16. c. Choice *b* can be left to Rembrandt's *Aristotle.*
17. c. See answer to question 3.
18. b. See answer to question 3 again.
19. c. According to the special theory of relativity, twins who experience different types of motion—one remains at rest while the other travels away and returns, for example—can have different ages.
20. a
21. c
22. d

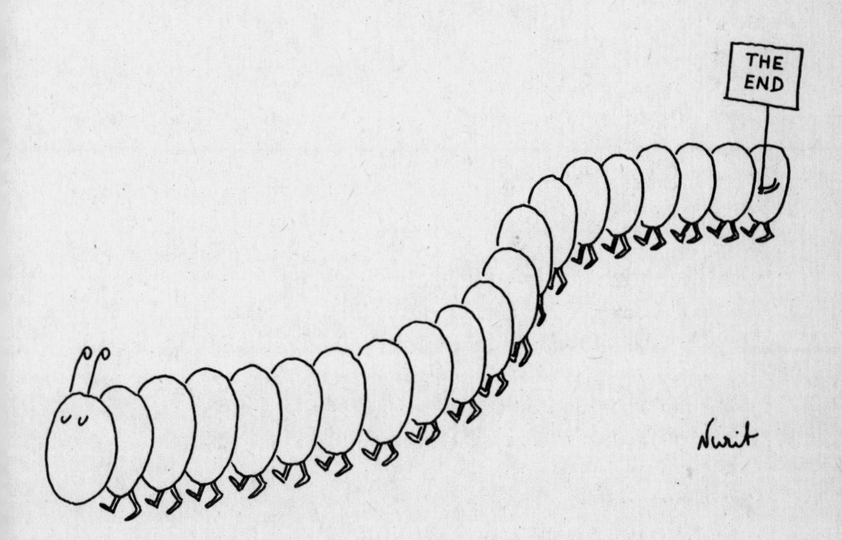

Photo Credits

Clara Barton, courtesy of American Red Cross.

Charles Darwin, reproduced by permission of The New York Academy of Sciences.

Sigmund Freud, reproduced by permission of the Archives of The New York Psychoanalytic Institute.

Jean Piaget, reproduced by permission of Howard E. Gruber.

Jonas E. Salk, reproduced by permission of Jonas E. Salk (photographer, D. K. Miller).

Benjamin Banneker, reproduced by permission of Johnson Publishing Co., Inc., Chicago, Illinois.

Andrei Sakharov (furnished by Charles Thompson) reproduced by permission of Jeri Laber, Helsinki Watch.

James D. Watson, reproduced by permission of James D. Watson.

T. D. Lee, reproduced by permission of T. D. Lee.

C. N. Yang, reproduced by permission of C. N. Yang.

George Washington Carver, painted by Laura Wheeler Waring, from Schomburg Center for Research in Black Culture, The New York Public Library, Astor, Lenox and Tilden Foundations.

Archimedes, engraving from Print Collection, Art, Prints and Photographs Division, The New York Public Library, Astor, Lenox and Tilden Foundations.

Photographs of Copernicus, Newton, Descartes, Galileo, and Einstein reproduced by permission of AIP Niels Bohr Library; that of Marie Curie, by the W. F. Meggers Collection; and that of Niels Bohr by the Margrethe Bohr Collection. Photographs courtesy of AIP Niels Bohr Library are those of Enrico Fermi, reproduced by permission of the University of Chicago, and of J. Robert Oppenheimer, by permission of the Los Alamos Scientific Laboratory.

Notes

Notes

Notes

Notes